A Primer to the Theory of Critical Phenomena

A Primer to the Theory of Critical Phenomena

Jurgen Honig
Purdue University
West Lafayette, IN, USA

Józef Spałek
Jagiellonian University
Kraków, Poland

ELSEVIER

Elsevier
Radarweg 29, PO Box 211, 1000 AE Amsterdam, Netherlands
The Boulevard, Langford Lane, Kidlington, Oxford OX5 1GB, United Kingdom
50 Hampshire Street, 5th Floor, Cambridge, MA 02139, United States

Notices

Knowledge and best practice in this field are constantly changing. As new research and experience broaden our understanding, changes in research methods, professional practices, or medical treatment may become necessary.

Practitioners and researchers must always rely on their own experience and knowledge in evaluating and using any information, methods, compounds, or experiments described herein. In using such information or methods they should be mindful of their own safety and the safety of others, including parties for whom they have a professional responsibility.

To the fullest extent of the law, neither the Publisher nor the authors, contributors, or editors, assume any liability for any injury and/or damage to persons or property as a matter of products liability, negligence or otherwise, or from any use or operation of any methods, products, instructions, or ideas contained in the material herein.

Library of Congress Cataloging-in-Publication Data
A catalog record for this book is available from the Library of Congress

British Library Cataloguing-in-Publication Data
A catalogue record for this book is available from the British Library

ISBN: 978-0-12-804685-2

Publisher: John Fedor
Acquisition Editor: Anita Koch
Editorial Project Manager: Katerina Zaliva
Production Project Manager: Maria Bernard
Designer: Maria Cruz

Typeset by VTeX

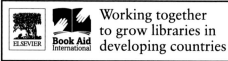

Contents

Preface

This book is intended to be an introduction to the basic theoretical background for the general area of critical phenomena taking place near a continuous phase transition. These phenomena encompass a singular behavior of physical properties, that is, the behavior of measurable physical quantities that may take infinite values at the critical point (e.g., at the critical temperature). Although many books and review articles already deal with this topic, they are often written at an advanced level of mathematical sophistication. We therefore attempted to provide the necessary background information readers will need to study the advanced presentations. Particular emphasis is placed on developing the concept of the order parameter and on a systematic approach starting from Landau mean-field theory.

We generally have tried to keep the discussion at a reasonably elementary level. Some familiarity with the elements of quantum mechanics, Fourier series and transforms, and complex variables is assumed. However, we have not hesitated to include some specialty topics relevant to our aim; this more advanced material, marked with (*), can be omitted on a first reading. Also, to render the individual chapters more self-contained, we have repeated some presentations in different chapters. Although we have tried to be careful, nothing can ever be expected to be error-free. We will therefore appreciate being informed where we have been remiss.

It remains to thank Elsevier personnel for their great patience in bringing this book to production. Particular thanks are due to Dr. Danuta Goc-Jagło for her skills in reconfiguring part of the original Word text into the LaTeX format and editing the figures.

This project was partly supported by The National Science Centre (NCN) through Grant MAESTRO, No. DEC-2012/04/A/ST3/00342.

Jurgen M. Honig
Józef Spałek
West Lafayette–Kraków, 2016–2017

Chapter 1

Introduction: Classical Phases and Critical Points

Phase transitions, and critical phenomena in particular, remain among the most spectacular properties in physics, materials science, and in other disciplines. At the transition materials in this category display mathematical singularities (infinities) in measurable physical quantities such as the specific heat, the bulk compressibility, or the magnetic susceptibility. What is equally important, when the critical point is approached these systems exhibit infinite fluctuations at all scales. At the same time, a surprising universality of these phenomena is observed, which are assembled into so-called universality classes. In this chapter, elementary experimental observations are briefly summarized to exemplify the principal characteristics of discontinuous vs. continuous classical phase transitions, i.e., those driven by thermodynamic fluctuations of the ordering process, as illustrated more precisely in later chapters.

1.1 PHYSICAL EXAMPLES: MACROSCOPIC PROPERTIES

We start with a macroscopic characterization of phase transitions. Everyday life is heavily dependent on the presence of water in its three manifestations as vapor, liquid, and solid. Each of these is referred to as a phase, a state of matter with uniform properties at equilibrium, that extend over a large region of space.[1] The phases are defined by their borders at which the measurable physical properties exhibit either singularities or discontinuities. The temperature T and pressure P can be independently varied over a wide range of conditions without affecting the overall appearance of each phase. However, there exist conditions where two of these phases do coexist, under circumstances where T and P can no longer be independently altered. This situation is displayed pictorially via a P–T phase diagram for water, as shown on a log–log scale in Fig. 1.1A and on a semilogarithmic scale in Fig. 1.1B. The curves represent the phase boundaries, i.e., the P–T dependence that must be met for the phase coexistence. Of special relevance to the entire topic of this book is the L-G branch that terminates at the so-called *critical point* CP with $T_c \simeq 647$ K and $P_c \simeq 218$ atm, where the menis-

1. The division into three phases is a matter of a well-tested convention, in which snow and clouds are excluded as separate phases.

A Primer to the Theory of Critical Phenomena. http://dx.doi.org/10.1016/B978-0-12-804685-2.00001-2

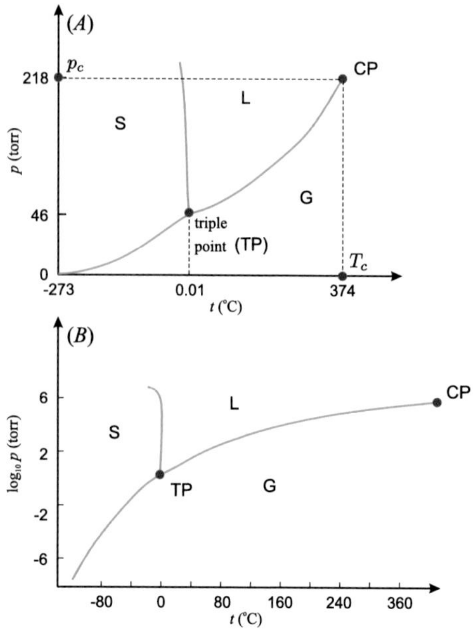

FIGURE 1.1 (A) Schematic phase diagram for water, depicting the range of stability of solid (S), liquid (L), and vapor (G). CP denotes the critical point; the triple point (TP) where the three phases coexist is also marked. This sketch serves as a template for phase diagrams for a huge variety of materials present under normal conditions. (B) Same phase diagram for water on a semilogarithmic scale.

cus separating liquid from vapor disappears entirely, indicative of the merging of liquid and vapor into a single entity termed a *critical fluid*. Just ahead of this event the system changes from a translucent to a cloudy (misty) appearance; this is due to almost uninhibited fluctuations in density that generate condensed clusters of size comparable to or larger than the wavelength of visible light, which scatter incident radiation. For completeness, we mention the *triple point* in the phase diagram ($T = 273.16$ K, $P = 6.03 \cdot 10^{-3}$ atm), at which all the three phases coexist.

Another indicator of the approach of water to critical conditions is the change in its latent heat of vaporization L as the temperature is raised. The enthalpy of vaporization $L = 2.25$ kJ/g at 273 K reflects the discontinuity in all physical properties of water as the liquid is converted to vapor. This process is generally referred to as a first-order (*discontinuous*) *phase transition*. Fig. 1.2 displays the progressive decrease of L with rising T that terminates close to $T_c = 647$ K, where the curve plunges to zero with infinite slope. Concomitantly, with rising temperature, the density of water diminishes, whereas that of steam rises until the two match at T_c. On crossing that temperature, the density of the sys-

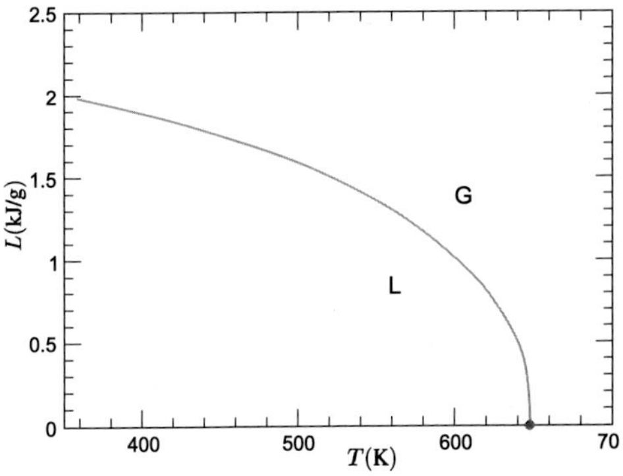

FIGURE 1.2 Latent heat of vaporization L of water as a function of temperature T at the critical pressure P_c. The end solid point marks the critical point at which the specific heat is singular, not just discontinuous as characterized by a latent heat L.

tem changes continuously, but other properties exhibit singularities, a process referred to as a *continuous phase transition*.[2] The precise manner of this coalescence is one instance of the more general study of continuous phase transitions in which we shall be interested at length.

A different hallmark of criticality is the existence of large spatial scale fluctuations. This reflects the inherent instability of the system close to criticality, to be documented in later chapters. Under these conditions, water and steam can be interconverted with negligible changes in the enthalpy. The process results in the presence of islands of water of widely variable size coexisting with steam as the critical point is approached. The constant bombardment of the condensed islands by gaseous water molecules gives rise to wild alterations in the local density of the system.

The above discussion obviously applies with appropriate changes to other systems. Fig. 1.1A then becomes a sort of generic phase diagram that displays the P–T dependence for a huge variety of materials, for which T_c/P_c takes values that range from 5.2 K/2.3 atm for He to 647 K/218 atm for water.

It may come as a surprise that completely different physical systems exhibit analogous if not exactly the same critical phenomena. Prominent among these are ferromagnetic materials. Fig. 1.3 displays the spontaneous magnetization M of metallic iron and nickel as a function of temperature T relative to that for

2. For historic reasons, this is also referred to as a second-order phase transition, but this designation has now fallen into disuse, as well as the term "first-order phase transition." Both of these terms were introduced by Paul Ehrenfest (1935) on phenomenological grounds. They have been replaced by the notion of discontinuous vs. continuous phase transitions introduced by Landau (1936), based on the concept of the order parameter and its changes at phase transition; these latter concepts will later be elaborated on in detail.

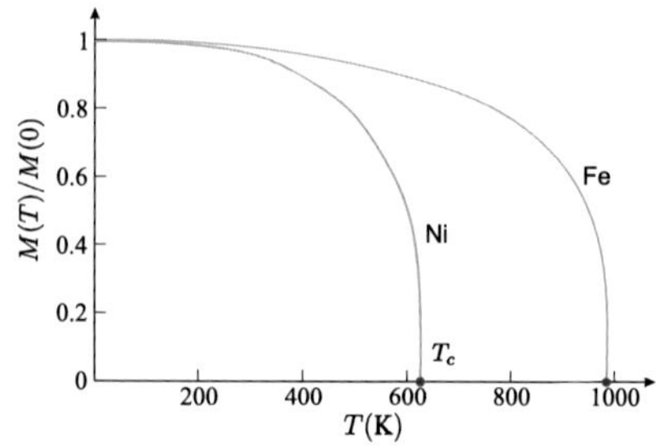

FIGURE 1.3 Schematic representation of the relative spontaneous magnetization of impure Fe and Ni as functions of temperature. Note the steep decline close to their respective Curie temperatures T_c with an infinite slope at that point.

$T \to 0$ K. Ignoring complications arising from magnetic domain structures, we note with rising T a continuous decrease in M, followed by a steep decrease with infinite slope at $M = 0$, i.e., at the Curie temperature $T = T_c$ (1043 K and 631 K for iron and nickel, respectively). Above this temperature, the material is in the paramagnetic, nanopolarized regime. The above reflects the gradual change with T in atomic spin alignments. Within a single domain, all equilibrated atomic moments are aligned at $T = 0$ K; with rising T, due to thermal fluctuations of the magnetic moment, dynamic islands of misaligned clusters make their appearance; as $T \to T_c$ these grow in size and number, thereby diminishing the overall magnetization.[3] Near T_c, these islands compete in size and prevalence with the original spin configuration; in fact, close to T_c, we encounter an enormous (infinite) range of size distributions of the continuously fluctuating aligned and random spin configurations, which then close in on the paramagnetic regime above T_c. It turns out that the theoretical analysis for the critical characteristics of both classes of materials is the same. This certainly should pique our interest. Later we introduce the so-called *universality classes* of critical behavior.

Another feature is worth noting: Fig. 1.4 shows a set of liquid–gas coexistence curves for eight distinct fluids near their respective critical points as plots of the scaled temperature T/T_c vs. the scaled density ρ/ρ_c The collapse of these data for materials with such widely varying properties onto a single curve is impressive, alerting us also to some kind of universality that calls for an explanation. Furthermore, close to T_c, the densities for the liquid and gas converge on each other according to the relation $|\rho - \rho_c| \sim |T_c - T|^\beta$ with $\beta \approx 0.326$, rather than with $\beta = 0.5$, as is predicted from the venerable van der Waals equa-

3. Strictly speaking, for $T \ll T_c$ collective wave excitations – spin waves (magnons) are the elementary excitations. Each of these carries a unit μ of the reversed magnetic moment. As T approaches T_c these gradually transform into misaligned clusters of variable size. At T_c it is difficult to identify such clusters, as they can be of any size.

TABLE 1.1 Values at critical $L \rightarrow G$ point for selected systems. In the last row, the characteristics of liquid ^{4}He at its selected boiling point are provided for comparison. ^{4}He is a quantum liquid

System	T_c (K)	P_c (atm)	ρ_c (g cm^3)
H_2O	647.5	218.5	0.325
CO_2	304.2	72.8	0.46
O_2	154.6	49.7	0.41
Xe	189.8	57.6	1.105
Ar	150.8	48.3	0.53
^{4}He	4.2	1	0.125

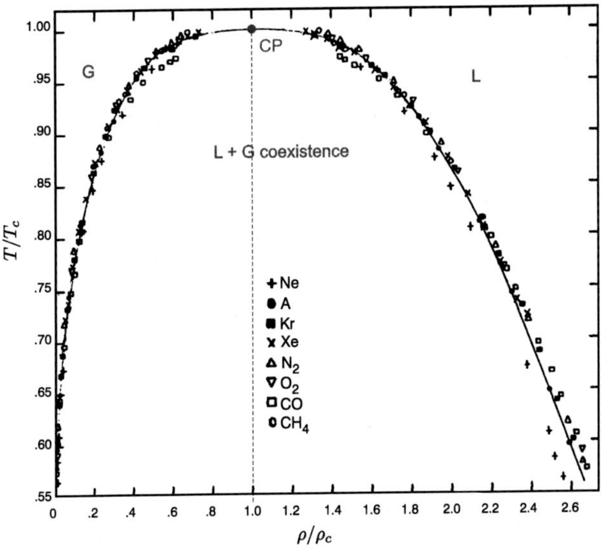

FIGURE 1.4 Plot of the relative temperature T/T_c vs. the scaled density ρ/ρ_c of eight different materials under conditions close to their critical temperatures for the liquid–gas transition. The gas phase is represented by the left part of the diagram. Note the collapse of all the data onto a single curve. [Adapted from E.A. Guggenheim, J. Chem. Phys. **13**, 253 (1945).]

tion of state. This discrepancy in β initially did not elicit much interest, but later became one of the catalysts that precipitated the entire study of critical phenomena. We should note that in Fig. 1.4 the relative pressure P/P_c is determined from the equation of state. The selected values (T_c, P_c, ρ_c) at the critical point are listed in Table 1.1. The universal plot obtained in ρ/ρ_c, T/T_c, and P/P_c is called the Guggenheim phase diagram, illustrating the so-called *principle of corresponding states* (Guggenheim, 1945).

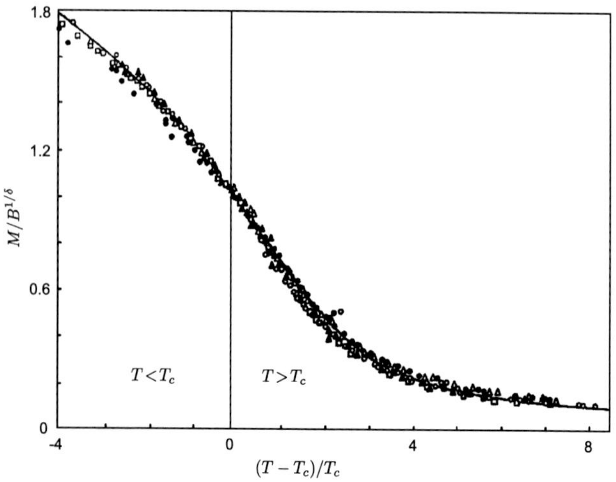

FIGURE 1.5 Experimental scaled spontaneous magnetization data (M divided by $B^{\frac{1}{\delta}}$) vs. scaled temperature across the critical point for $CrBr_3$, EuO, Ni, YIG, and Pd_3Fe. The collapse of the data into a single curve is noteworthy [after H.E. Stanley, Rev. Mod. Phys. **71**, S358 (1999)].

An analogous situation obtains for magnetic materials. In Fig. 1.5 we show curves depicting the variation of the scaled magnetization $M/B^{1/\delta}$ vs. $(T - T_c)/T_c$ for five different materials of widely different magnetic characteristics. The data collapse effect is apparent; note the asymptotic descent to the paramagnetic regime ($M \to 0$) for temperatures well above T_c.

1.2 SINGULARITY OF THE SPECIFIC HEAT

To demonstrate explicitly the divergence of the physical quantities at a continuous phase transition, we have plotted in Fig. 1.6A the overall temperature dependence of the specific heat for a normal-liquid to superfluid transition in condensed ^{4}He. The singularity appears for $T = T_\lambda \simeq 2.17$ K. This transition is called the λ transition in liquid ^{4}He since the data resemble the shape of the Greek letter λ. To show explicitly that this dependence exhibits a mathematical singularity, we have replotted in Figs. 1.6B–D the same dependence as a function of $T - T_\lambda$ for gradually narrower ranges in temperature close to T_λ. We can clearly see that, even at the microKelvin scale, the data exhibit the same type of singularity, namely a $\ln|T - T_c|$ dependence[4] with an adjustable constant of proportionality.

4. Strictly speaking, the function is $A_\pm - B\ln|T - T_c|$, where $A_+ = -0.65$ for $T > T_\lambda$, $A_- = 4.55$ for $T < T_\lambda$, and $B = -3.00$ in units of J/mol K (from M.I. Buckingham and W.M. Fairbank, Prog. Low Temp. Phys. **3**, 80 (1961)).

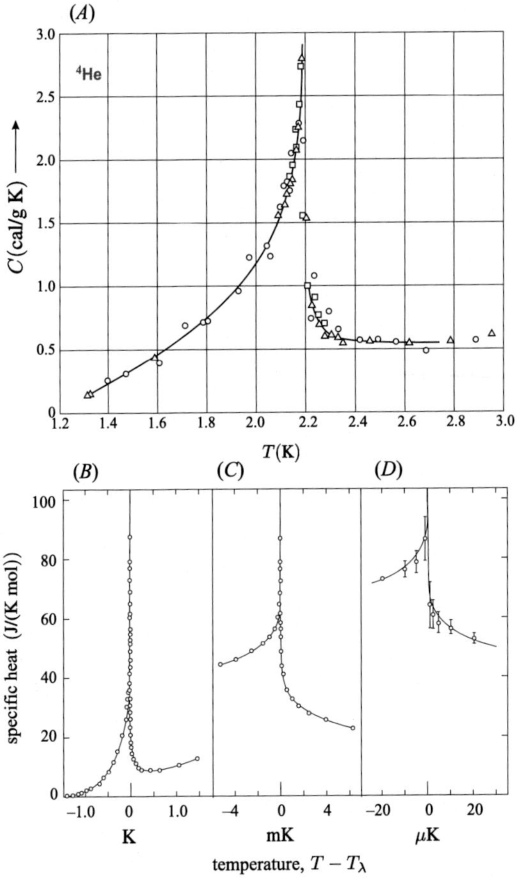

FIGURE 1.6 (A) Overall temperature dependence of the specific heat near the normal-superfluid transition in condensed ^{4}He (original data of Keesom group from Leiden, 1930s). (B)–(D) The same dependence around T_λ on Kelvin (B), miliKelvin (C), and microKelvin (D) scales. The continuous curve which represents the divergent $\ln|T - T_\lambda|$ behavior is the same in each case. [Adapted from J.A. Lipa et al., Phys. Rev. Lett. **76**, 944 (1996).]

To compare the above behavior of the specific heat to that observed in a discontinuous phase transition, we show in Fig. 1.7 the temperature dependence of the specific heat at constant pressure (C_p) for pure magnetite Fe_3O_4 near the *Verwey transition* taking place at $T = T_v \simeq 120$ K. We see a clear sign of a spike representing the latent heat of the transition over a quasicontinuous background. No sign of the λ-type behavior can be seen. Such behavior is a landmark of a discontinuous transition.

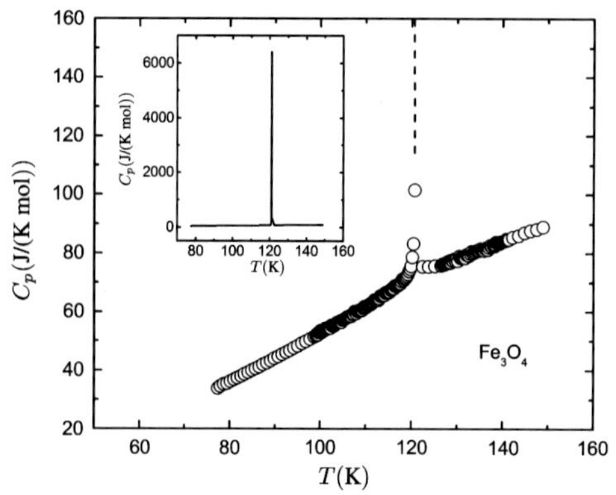

FIGURE 1.7 Temperature dependence of the specific heat (C_p) at constant pressure near the discontinuous Verwey (semiconductor–semiconductor) transition in pure magnetite Fe_3O_4. The spikes (cf. also inset) represent the latent heat, illustrating the discontinuous character of the transition. No sign of λ-type transition is detected in this case [from A. Kozłowski et al., Phys. Rev. B **54**, 12093 (1996)].

1.3 REMARKS

The study of phase transitions in general, and continuous ones in particular, involves phenomena well beyond those in the laboratory. Indeed, contemporary theories attribute the existence of our observable universe to a phase change that is generated from a preexisting singular vacuum. Moreover, contemporary theories model processes that deal with continuous phase transitions in quantum field theories and in astrophysical objects such as neutron stars. Thus, the above discussion represents just a tiny fraction of interesting phenomena, both experimental and theoretical, displayed near the critical point of a system. Aside from the large body of experiments that call for an explanation, there is, of course, the purely intellectual challenge of developing the appropriate theoretical models of those singular phenomena. These efforts are summarized in a number of monographs listed at the end of the book.

The current volume represents an attempt to prepare readers for the in-depth study required to master that material.

Chapter 2

The Ising Model and Its Basic Characteristics in the Mean Field Approximation

In this chapter, we analyze probably the simplest microscopic model exhibiting a continuous phase transition and its fundamental characteristics: the concept of the self-consistent field, the order parameter, and the resulting self-consistent equation. From these basic concepts there follows the entire statistical thermodynamics of the system, yielding the temperature and field dependence of measurable physical quantities, such as the magnetization, the magnetic susceptibility, and the specific heat. Emphasis is placed on analytical discussions; so we often resort to an approximate analysis, e.g., when the temperature is close to the critical temperature T_c for the phase transition in question. A full numerical discussion is deferred, as a specific task for the student, to the end of the chapter.

2.1 THE ISING MODEL AND ITS HAMILTONIAN

As we set out on our theory journey, to study phase transitions and critical phenomena, consider first a simple example, the so-called Ising (pr. Eezing) model, consisting of a regularly spaced two-dimensional lattice array of N elementary units (spins), placed in an applied magnetic field B oriented in the perpendicular direction.[1] Each lattice constituent is associated with a magnetic dipole whose spin $S_i = -S, -S+1, \ldots, +S$ is oriented along with, or opposite to, the applied field, with the corresponding energy $-\mu S B$. We seek to investigate the critical properties of this elementary system.

The Hamiltonian (energy) $\mathcal{H}$ for such a configuration is specified by

$$\mathcal{H} = -J \sum_{\langle i,j \rangle} S_i S_j - \mu \sum_i S_i B, \tag{2.1}$$

1. Quantum mechanically, S_i is the spin z-component S_i^z of the quantum particle or of an entire atomic shell of electrons with parallel spins, due to Hund's rule. Throughout the following discussion, we thus set $S_i \equiv S_i^z$. Also, $\mu \equiv g\mu_B$, where g is the Landé factor ($g = 2$ for the electron), and $\mu_B \simeq 9.2732 \cdot 10^{-24}\ \mathrm{J\,T^{-1}} = 9.2732 \cdot 10^{-21}\ \mathrm{erg\,G\,s^{-1}}$, is the Bohr magneton.

A Primer to the Theory of Critical Phenomena. http://dx.doi.org/10.1016/B978-0-12-804685-2.00002-4

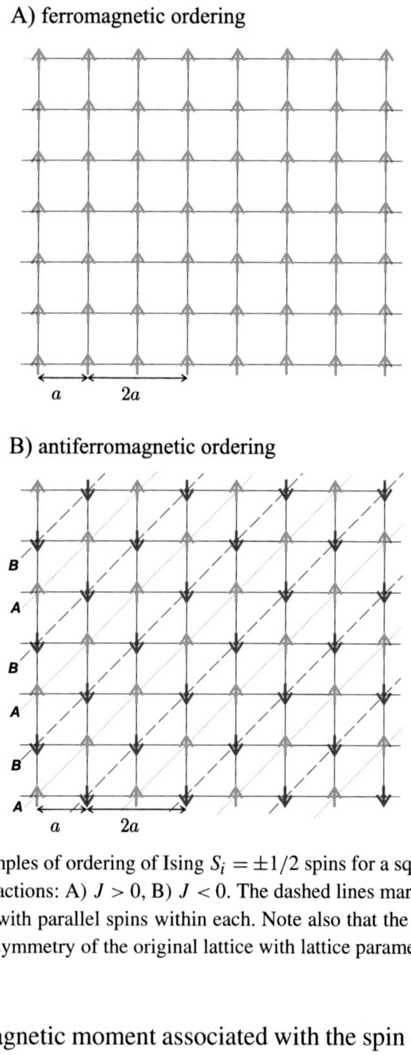

FIGURE 2.1 Two examples of ordering of Ising $S_i = \pm 1/2$ spins for a square lattice with nearest-neighbor exchange interactions: A) $J > 0$, B) $J < 0$. The dashed lines mark the sublattices for case B), i.e., the subsystems with parallel spins within each. Note also that the spin ordering in case B) breaks the translational symmetry of the original lattice with lattice parameter a.

where μS_i is the magnetic moment associated with the spin on site i, J is a positive coupling constant (exchange integral) for interactions between z nearest neighbors, and the symbol $\langle i, j \rangle$ restricts the summation to nearest-neighboring sites j of each i and vice versa[2]; longer-range interactions are ignored. Two simple types of ordering derived from Eq. (2.1) for $B = 0$ are drawn schematically in Fig. 2.1.

We now introduce further approximations. First, write $S_i \equiv \langle S_i \rangle + (S_i - \langle S_i \rangle)$, where the angular brackets signify expectation values to be defined later. By definition, we set $\langle S_i \rangle \equiv M$ as the uniform average spin (density),

2. The summation is thus performed for $i, j = 1, \ldots, N$ and $i \neq j$.

also generally referred to as the magnetization,[3] a designation we follow here. $(S_i - \langle S_i \rangle) \equiv \delta S_i$ represents the (presumably small) fluctuating deviations of the local spin from its average value. Taking the view that those deviations can be (for the time being) neglected, Eq. (2.1) may be recast in the following form:

$$\mathcal{H} = -\mu \sum_i S_i \left(B + \frac{J}{\mu} \sum_{j(i)} S_j \right) \approx -g\mu_B \sum_i S_i \left(B + \frac{Jz}{g\mu_B} \langle S_j \rangle \right), \quad (2.2)$$

where we have replaced the total field (the second term) involving the interactions with z nearest neighbors by its average value. This is the so-called *mean-field approximation*. It amounts to stating that the summed effect of the neighbors is averaged out. For such an approximate form of the system energy, we determine the statistical thermodynamics, i.e., the temperature T and applied magnetic field B dependences of relevant quantities. But first we have to determine the *ad hoc* average $\langle S_i \rangle = M$.

2.2 THE ISING PARTITION FUNCTION AND FREE ENERGY[4]

We sum over the N identical contributions and define an effective field as $B_{\text{eff}} \equiv JzM + \mu B$, whereby the Hamiltonian assumes the form

$$\mathcal{H} = -\sum_i S_i B_{\text{eff}}. \quad (2.3)$$

With[5] $S_i = \pm S = \pm\frac{1}{2}$, we can then specify the partition function (statistical sum) for the system as ($\beta \equiv \frac{1}{k_B T}$, where k_B is the Boltzmann constant)

$$\begin{aligned} \mathcal{Z} \equiv \operatorname{Tr} e^{(-\beta\mathcal{H})} &\equiv \sum_{S=S_i} e^{-\beta \sum_{i=1}^{N} B_{\text{eff}} S_i} = \prod_{i=1}^{N} \left(\sum_{S=S_i} e^{-\beta B_{\text{eff}} S_i} \right) \\ &= \left(e^{\beta S B_{\text{eff}}} + e^{-\beta S B_{\text{eff}}} \right)^N = (2\cosh \beta S B_{\text{eff}})^N, \end{aligned} \quad (2.4)$$

3. Magnetization is physically defined as $\mu\langle S_i \rangle$; then it signifies the magnetic moment per lattice site (e.g., per atom). Here the minus sign is ignored, so $\boldsymbol{S}_i$ and $\boldsymbol{M}$ are parallel; thus, by "spin" we designate the magnetic moment or magnetization per atomic site in dimensionless units.

4. For general properties and the meaning of the partition function, see Appendix 2.A. For completeness the principles and some detailed results are assembled in the Appendices.

5. This approximation is valid only for $S = \frac{1}{2}$. In the literature one frequently encounters the use of $S = 1$. This is admissible if one is interested only in the development of concepts. However, for the analysis of actual physical situations the selection $S = 1$, involves three components of $S_i \equiv S_i^z = 0, \pm 1$, in the construction of the partition function Z.

which, in turn, leads to the so-called *Landau free energy* $\mathcal{F}$ for the magnetic properties, which is specified by $\mathcal{F} = -k_B T \ln \mathcal{Z}$; thus,

$$\mathcal{F} \equiv \mathcal{F}(T, B, M) = -Nk_B T \ln(2\cosh \beta S B_{\text{eff}}). \tag{2.5}$$

Parenthetically, even though the quantity B_{eff} represents the total magnetic field acting on each of the magnetic moments, $\mathcal{F}$ is a function of both B and M simultaneously. Because these two quantities appear concurrently as *conjugate variables* in a function of state, $\mathcal{F}$ as specified by Eq. (2.5) cannot be called the Helmholtz free energy (see Appendix 2.A). The relation between the Landau free energy (2.5) and the Helmholtz function will be discussed later.

Consider first the zero-field case ($B = 0$) and set $u \equiv \beta J z M S$. Then expand the hyperbolic function as an ascending power series in u up to u^4, as shown in Appendix 2.B; this restricts us to small departures from $u = 0$. We also combine the $\ln 2$ factor with the free energy of the system, exclusive of the magnetic contributions. We then find that

$$\mathcal{F} \approx F_0(T) - Nk_B T \left\{ \frac{u^2}{2} - \frac{u^4}{12} \right\}, \tag{2.6}$$

where $F_0(T) \equiv -Nk_B T \ln 2$. Since JzS^2 has the dimension of energy, we may associate this product with a characteristic temperature $k_B T_c = JzS^2$, which simplifies the subsequent analysis and also has the fundamental significance shown below. Inserting the proper expression for u, we end up with the relation

$$\mathcal{F}(T, M) \approx F_0 + NJzS^2 \left[\left(1 - \frac{T_c}{T}\right) \frac{M^2}{2S^2} + \left(\frac{T_c}{T}\right)^3 \frac{M^4}{12S^4} \right] + o(M^6). \tag{2.7}$$

This expression contains a number of experimental parameters, including the temperature T. As stated earlier, $\mathcal{F}$ has the form of a free energy function but is not the thermodynamic free energy because it has not been subjected as yet to an equilibrium constraint.[6] We attend to the equilibration shortly.

The examination of Eq. (2.7) is divided into three parts: On introducing the quantity $m \equiv \frac{M}{S}$ we note that when $T > T_c$, both terms in the square brackets are positive, so that $\mathcal{F} - F_0$ rises with m from zero, as sketched in the top curve of Fig. 2.2. Precisely, at $T = T_c$, the free energy density varies as m^4, emerging from zero as the somewhat flattened curve depicted in the middle. When $T < T_c$ the first term in brackets is negative; initially, $\mathcal{F} - F_0$ decreases with rising $|m|$, but the positive contribution from m^4 ultimately outweighs the negative one, giving rise to the bottom curve in Fig. 2.2. Now the minimum of the free energy

6. In the functions of state, only one of those variables should appear (see Appendix 2.A for details). In the remaining part of the chapter, we discuss how to transform $\mathcal{F}$ into the physical free energy.

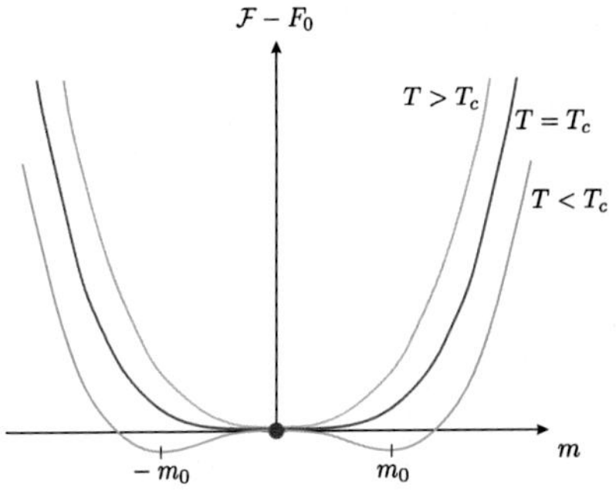

FIGURE 2.2 Schematic representations of the Landau free energy as a function of m (order parameter) in the three indicated cases. Note that the temperature $T = T_c$ separates ordered states (for $T < T_c$, $\langle S_i \rangle \neq 0$) from the disordered state ($T > T_c$, $\langle S_i \rangle = 0$). This is the reason why T_c is called the *critical temperature*. The full meaning of T_c is elaborated on in the main text.

no longer occurs at $m = 0$, but is shifted to $\pm m_0$. Later, when dealing with critical phenomena, we will again encounter this three-pronged variation.

2.3 EQUILIBRIUM CONDITIONS AND THERMODYNAMICS

At this point, we introduce the equilibrium constraint on the function (2.5). We eliminate the thermodynamic quantity M from $\mathcal{F}(T, B, M)$ by imposing the condition that the physical (Helmholtz) free energy is realized as a minimum of $\mathcal{F}(T, B, M)$. This means that we set $\frac{\partial \mathcal{F}}{\partial M}|_{M_0} = 0$ and $\frac{\partial^2 \mathcal{F}}{\partial^2 M}|_{M_0} > 0$, which leads to the self-consistent equation for M in the form

$$M = S \tanh u = S \tanh(\beta J z S M). \tag{2.8}$$

This algebraic equation clearly shows that M and T now are no longer independent variables. In our current approximation scheme, the hyperbolic function reads $\tanh x \simeq x - x^3/3$, and Eq. (2.8) assumes the form

$$M = \beta J z S^2 M - \frac{1}{3}(\beta J z)^3 S^4 M^3 + o(M^5), \tag{2.9}$$

whence

$$1 \simeq \beta J z S^2 - \frac{1}{3}\frac{(\beta J z S^2)^3 M^2}{S^2}, \text{ i.e., } M \equiv M_0 = \pm\left[3\frac{1 - \beta J z S^2}{(\beta J z)^3 S^4}\right]^{1/2}. \tag{2.10}$$

Of special interest is the ferromagnet precisely at the critical temperature $T = T_c$, $M = 0$, where

$$k_B T_c = J z S^2. \tag{2.11}$$

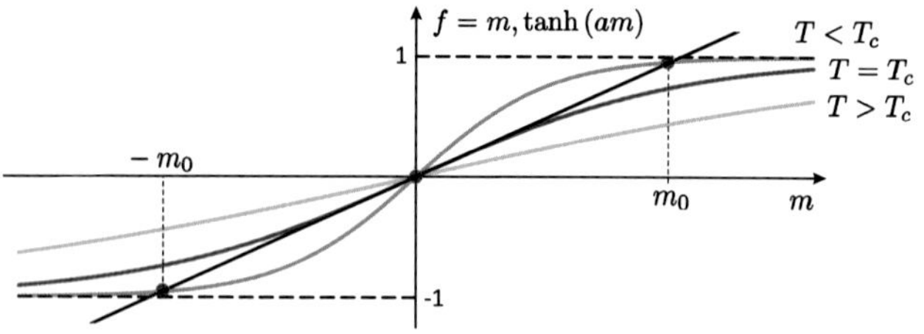

FIGURE 2.3 The graphical illustration of the self-consistent equation (2.12). The straight (black) line represents the l.h.s. $y = x = m$, whereas the r.h.s. is represented by $y = \tanh(am)$ with $a \equiv T/T_c$. We have two stable solutions $m = \pm m_0$ for $T < T_c$ $(a < 1)$ and a single one $m \equiv 0$ for $T \geq T_c$ $(a \geq 1)$, since in the latter case the two curves have only a common point at $x = 0$. This figure has the same interpretation as Fig. 2.2.

Also, by defining $m \equiv M/S$ we can rewrite (2.8) in the universal form

$$m = \tanh\left(\frac{T_c}{T} m\right), \tag{2.12}$$

where T_c is indeed the only parameter determining the behavior of $m = m(T)$. We have thereby identified T_c as a characteristic temperature. Moreover, in the neighborhood of $M = 0$, T does not differ substantially from T_c. At $T = T_c$, we see that this equation takes the form $m = \tanh m$, with the only admissible solution $m = 0$. Note further than Eq. (2.12) does not admit a physically acceptable solution for m when $T > T_c$. The three situations can be represented graphically as shown in Fig. 2.3.

Beginning with Eq. (2.7), and close to the critical temperature, where Eq. (2.B.5) in Appendix 2.B applies, we end up with the basic expression

$$\mathcal{F}(T) = F_0 + NJzS^2\left[\left(\frac{T}{T_c} - 1\right)\frac{M^2(T)}{2S^2} + \frac{M^4(T)}{12S^4}\right] + o(M^6), \tag{2.13}$$

which indicates that the thermodynamic free energy density involves solely T as the independent variable (if $B = 0$). The numerical solution of the full equation (2.8) reflects the features just discussed and the condition $|M| \leq S$. For $T \to 0$, $\tanh(...) \to 1$ and $M \to S$ for every lattice site i. This type of ordering has been drawn schematically in Fig. 2.4 (see also Problem 2.1 for details). The variation of m with T is depicted in Fig. 2.4, showing the restriction of the equilibrium values of the M/S ratio to the range $0 \leq m \leq 1$. This leads to the introduction of a new construct, the so-called *order parameter*, whose properties will be more fully discussed later. In the present context $m = 1$ or 0 correspond respectively to complete ferromagnetic ordering at $T = 0$ (the so-called spin ordered regime) and to total disorder at high temperatures (the so-called spin disordered

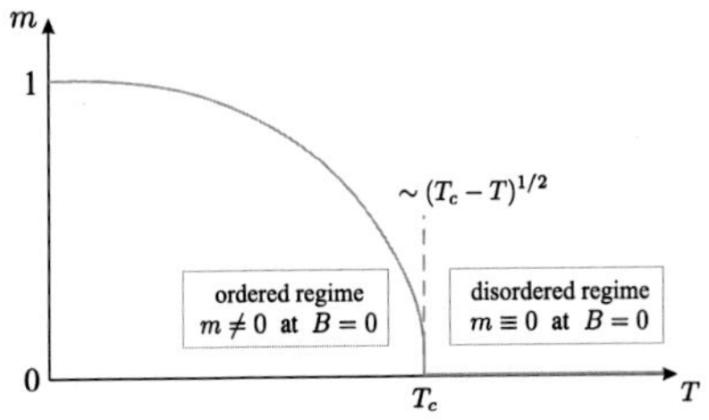

FIGURE 2.4 Schematic representation of the temperature dependence of the magnetic moment regarded as the order parameter. Note the singular behavior of dm/dT at $T = T_c$.

regime), of the magnetic moments. Thus, $m(T)$ represents the average intermediate degree of alignment at thermodynamic equilibrium of these moments at temperature T, hence, the average scaled magnetization M/S at each site. The transition from $m = 1$ to $m = 0$ is continuous, but involves a singularity as $T \to T_c$.

Parenthetically, expression (2.13) is termed the Landau free energy if the higher-order terms are neglected.

2.4 ELEMENTARY CONSEQUENCES OF THE MODEL

We can do more. In what follows, we further discuss important details.

(i) Let us first deal with the spontaneous magnetization at $T < T_c$ in the absence of a magnetic field. Referring back to Eq. (2.11), rewrite Eq. (2.12) for $m \neq 0$ as

$$1 = \frac{T_c}{T} - \frac{1}{3}\left(\frac{T_c}{T}\right)^3 m^2. \tag{2.14}$$

Now, introduce the dimensionless quantity

$$t \equiv \frac{T}{T_c} - 1, \tag{2.15}$$

which renders $t \ll 1$ near the critical temperature. Then, to the leading order in t, Eq. (2.12) reads

$$m = \left[-3\left(\frac{T}{T_c}\right)^3 t\right]^{1/2} = \left[-3(1+t)^3 t\right]^{\frac{1}{2}} \approx \sqrt{-3t} = \sqrt{3\frac{(T_c - T)}{T_c}}. \tag{2.16}$$

Note several points. Close to T_c, we neglect terms in t^2 or higher powers. Then, the magnetization varies as the square root with deviations from the critical point, when approached from below. We may determine the variation of m with arbitrary T by solving Eqs. (2.5) and (2.8) numerically; the resulting curve is sketched in Fig. 2.4 (see also Problem 2.1). This provides one measure of the inadequacy of the present model: the magnetization near the critical temperature goes to zero as the square root of $T_c - T$; experimentally, as shown in Fig. 1.5, the limiting value of zero is also reached with infinite slope, but with the critical behavior $M \sim t^{\beta}$, where $\beta \approx \frac{1}{3}$.

(ii) Of interest also is the change in magnetic properties in a weak magnetic field at temperatures close to the critical temperature. Here one must recognize that the experimental measurements involve the magnetization M rather than the scaled quantity m. Accordingly, we modify Eq. (2.8) which now reads $M = \tanh(\beta J z M + \beta \mu B)$. We also retain only the first order terms. Then:

(a) for $T > T_c$, the magnetization arises solely from the existence of the field. This is the case because

$$M = \beta J z M + \beta \mu B = \frac{T_c}{T} M + \beta \mu B, \tag{2.17}$$

and for $B = 0$, we encounter only the physical solution $M = 0$; otherwise, we would in general obtain the incorrect relation $\beta J z = 1$. Equation (2.17) may be rewritten in the form

$$M = \frac{\mu B}{\left(1 - \frac{T_c}{T}\right) k_B T} = \frac{\mu B}{k_B (T - T_c)}, \tag{2.18}$$

showing that in a weak applied magnetic field the magnetization diverges as the critical temperature is approached from above for any $B \neq 0$. To determine the magnetic susceptibility χ_T, note that the total magnetization of the system is specified by $N \mu M$, whence the (isothermal) magnetic susceptibility is

$$\chi_T \equiv \left(\frac{\partial N \mu^2 B}{\partial B} \right) \bigg|_{B \to 0} = \frac{N \mu^2}{k_B (T - T_c)}. \tag{2.19}$$

Note that this experimentally measurable quantity is infinite at $T = T_c$! This is the reason why T_c is called the critical temperature.

(b) Below, but close to T_c, we must include the additive effect of the spontaneous magnetization. We assume that the result should not deviate significantly from the above magnetization $M_0 = \sqrt{-3t}$, so that, in the presence of $B \neq 0$, we may write

$$M = M_0 + \delta M, \tag{2.20}$$

where δM is a small quantity (B is regarded as small). The subsequent operation is simplified by adapting Eq. (2.8) to read $M = \tanh(\beta J z M + \beta \mu B)$ and then carrying out the inverse operation

$$\tanh^{-1} M = \beta J z M + \beta \mu B \approx M + \frac{1}{3}M^3, \tag{2.21}$$

where we have retained only the first two terms in the expansion of the left. Effectively, we can rewrite the above in the form

$$\left(\frac{T_c}{T} - 1\right) M = \frac{1}{3}M^3 - \frac{\mu B}{k_B T}. \tag{2.22}$$

As shown in Appendix 2.B, on introducing appropriate approximations and using Eq. (2.20) with $M_0^2 = -3t$, we end up with the relation

$$\delta M = \frac{-\mu B}{2k_B(T - T_c)}, \tag{2.23}$$

whence the magnetic susceptibility is given by

$$\chi_T = \left(\frac{\partial N\mu\delta M}{\partial B}\right)_{B\to 0} = \frac{N\mu^2}{2k_B(T_c - T)} \sim (T_c - T)^{-1}. \tag{2.24}$$

Thus, we obtain the same divergent power law as in case (a), but differing in sign and by an additional numerical factor of $\frac{1}{2}$.

This and the previous divergence of χ_T has a deep significance. Namely, the magnetic response, as characterized by the field-induced magnetization to the very weak applied field, is infinite at $T \to T_c \pm 0$. This situation corresponds to the presence of infinite fluctuations (changes) at the critical point. This circumstance is a characteristic feature of all systems near continuous phase transitions.

(iii) Of additional interest is the dependence of the magnetization on the applied magnetic field exactly at the critical temperature. Return to Eq. (2.21) and set $T = T_c$, whence

$$M = M + \frac{1}{3}M^3 - \frac{\mu B}{k_B T_c}, \tag{2.25}$$

or

$$B = \frac{k_B T_c}{3\mu} M^3, \tag{2.26}$$

so that the magnetization at $T = T_c$ obeys the proportionality $M \sim B^{\frac{1}{3}}$.

(iv) Finally, we investigate the heat capacity in the absence of an applied field as we pass the critical temperature. Just below T_c, the magnetic contribution to the internal energy is given by

$$U(T) = -\frac{NJz}{2}M_0^2 \approx -\frac{3}{2}NJz\left(\frac{T_c - T}{T_c}\right) = \frac{3}{2}Nk_B(T - T_c). \tag{2.27}$$

Now introduce the approximate relation $F(T) = U(T) - TS \equiv F_0(T) + U(T)$, so that $U(T) = F(T) - F_0(T)$. Then the heat capacity just below T_c is specified by

$$C = \frac{3Nk_B}{2}. \tag{2.28}$$

Above T_c, $U(T) = 0$; hence, in the Ising model, the heat capacity exhibits a discontinuity on crossing the critical temperature.

A more rigorous derivation of some of the above results will be provided in the next chapters.

2.5 DISCUSSION: CRITICAL EXPONENTS

It is rather impressive how many physical predictions can be obtained from such a rudimentary model as the Ising lattice. However, we cannot expect the above results to provide an accurate interpretation of experimental observations; the discrepancies will be discussed in due time. An important reason for the failure of the primitive model is our neglect of the fluctuations δS_i in Eq. (2.2); every spin S_i has been replaced by its average value S, a procedure that is known as the *mean field approximation*. Also, the only allowed configuration is one in which the spins are in complete (at $T = 0$) or partial (at $0 < T < T_c$) alignment even without an applied magnetic field.[7] To conform more closely to reality, we should adopt the Heisenberg model in which each spin is represented as a vector whose specification requires not only the magnitude but also the spin orientation. Thus, at best, we have devised a highly approximate approach to deal with the experimental observations in real-life situations.

A preliminary comparison may be instructive: For dealing with experimental results, the following standard notation has been in use: At, or close to, the critical temperature:

- the heat capacity is specified by

$$C \sim |T - T_c|^{-\alpha}; \tag{2.29}$$

7. We can intuitively think that the spontaneous order for $T < T_c$ appears when we consider the system in an external field $B \neq 0$, and subsequently go to the limit $B = 0$ in a continuous manner. Such an interpretation is, however, not entirely general, since we obtain the solution $M_0 \neq 0$ by solving directly Eq. (2.8) or (2.12). We see that the term JzM/μ plays the role of such a (self-consistent) field. Therefore, it is correct to state that at T_c the degeneracy of the states $S_i = \pm S$ is broken even for $B = 0$, i.e., there is a spontaneous breakdown of symmetry of the degenerate states, and a spontaneous order of the spins appears at $T < T_c$. In other words, the spontaneous order represents an emergent phenomenon of Nature rather than that induced by the presence of a vanishing external perturbation.

TABLE 2.1 Examples of critical exponents for d-dimensional lattices

Exponents	Mean field result	$d = 2$	$d = 3$ (numerical calculations)
α	discontinuous	$\ln \lvert T - T_c \rvert$	0.110
β	0.5	0.125	0.312
γ	1	1.75	1.238
δ	3	15	5.0

- the average spin (or magnetization) is specified by

$$M_0 \sim (T_c - T)^{\beta}, \ T < T_c; \tag{2.30}$$

- the isothermal magnetic susceptibility is specified by

$$\chi_T \sim |T - T_c|^{-\gamma}; \tag{2.31}$$

- the relation between the applied magnetic field and the average spin (or magnetization) is specified by

$$B \sim M^{\delta}, \ T = T_c. \tag{2.32}$$

The four exponents listed are known as *critical indices* or *critical exponents*. It is rather unfortunate that β as used above may be confused with $\beta = \frac{1}{k_B T}$. We must sort out their meaning within the context of their use. It is of note that, as $T \to T_c$, all the above physically measurable quantities display such power law behavior, i.e., with only one exponent, albeit a real number.

It is of interest to compare in Table 2.1 the above exponents with those obtained by a more involved analysis for the case $d = 2$ and with numerical calculations for simple cubic Ising lattices.

The uncertainties in numerical values are in the next digits. Table 2.1 uncovers striking differences: aside from numerical discrepancies, the mean field values do not depend on the spatial dimensionality of the system, whereas the sophisticated calculations do. This is a feature that will occupy us at length in our later development. At this point, it is appropriate to mention that the two-dimensional case $d = 2$ was solved rigorously by L. Onsager in 1944 in what is justifiably considered to be a real *tour de force*, for which he was awarded the Nobel Prize in 1968. His derivation leads to the result $M_0 \sim (T_c - T)^{\frac{1}{8}}$, as compared to $\beta = \frac{1}{2}$ in the present model. No exact solution is known for three dimensions, but the discrepancy between the results based on the Ising model and the corresponding numerical simulations for three dimensions is still substantial. As anticipated on physical grounds, the predictions of the Ising model improve with increasing dimensions because the greater number of nearest-neighbor interactions tends to average out fluctuations. Actually, the exact results for a sufficiently high-dimensional lattice are the same as those obtained from the mean field approach. A more detailed comparison between theory and experiment will be provided later.

2.6 THE MEANING OF THE ISING MODEL AND MEAN FIELD THEORY

As stated earlier, the mean field theory, or rather, *mean field approximation* provides the simplest description of the mutual coupling between the neighbors, as represented above by discrete (lattice) spins and their corresponding discrete values. The mean field approximation for the true three-dimensional spins reduces to that for the Ising spins as long as the resultant spin ordering is collinear (see Fig. 2.1) since then the transverse spin components can be disregarded. Nonetheless, the fluctuations are completely different: in the latter situation, the rotational degrees of freedom do matter in a basic manner.

The Ising model is applied well beyond the statistical or condensed-matter physics since it is a prototype for systems coupled by discrete degrees of freedom. Also, it is very useful in the analysis of magnetic systems with strong uniaxial anisotropy of the spin order. Some of these problems will be discussed in what follows.

2.7 PROBLEM 2.1: NUMERICAL CHECKOUT

Solve Eq. (2.12) by using the equation form

$$m = \tanh\left(\frac{m}{a}\right) \quad \text{with} \quad 0 \le a \equiv T/T_c \le 1. \tag{2.33}$$

Check that the results are the same as those displayed in Table 2.2. In particular, check that near $T = T_c$ ($t = 1 - T/T_c \ll 1$), a good approximation of the solution is

$$m = \pm\sqrt{3a^2(1-a)} = \pm\sqrt{3\left(\frac{T}{T_c}\right)^2\left(1 - \frac{T}{T_c}\right)}. \tag{2.34}$$

Plot the curves of $m = m(T)$ according to (2.33) and (2.34) and check that they closely agree for $a = T/T_c \gtrsim 0.9$ (i.e., $t \lesssim 0.1$), as shown in Fig. 2.5. Note that $(dm/dT)|_{T\to T_c} \to \infty$, whereas $(dm/dT)|_{T\to 0} \to 0$. Also, since $0 \le |m| \le 1$, we can regard m as the order parameter in the classical thermodynamic sense.

Usually, one can neglect the factor $(T/T_c)^2$ in Eq. (2.34) as $T \to T_c$ (see Fig. 2.5).

TABLE 2.2 Numerical values of the temperature dependence of the order parameter. The fifth column contains the values of *m* obtained from Eq. (2.34)

$a = T/T_c$	$m = M/S$	$a = T/T_c$	$m = M/S$	m
0	1	0.65	0.872	0.666
0.05	1.000	0.7	0.829	0.664
0.1	1.000	0.75	0.776	0.650
0.15	1.000	0.8	0.710	0.620
0.2	1.000	0.85	0.630	0.570
0.25	0.999	0.9	0.525	0.493
0.3	0.997	0.95	0.379	0.368
0.35	0.993	0.96	0.341	0.333
0.4	0.986	0.97	0.296	0.291
0.45	0.974	0.98	0.243	0.240
0.5	0.958	0.99	0.173	0.171
0.55	0.936	0.995	0.122	0.122
0.6	0.907	1	0	0

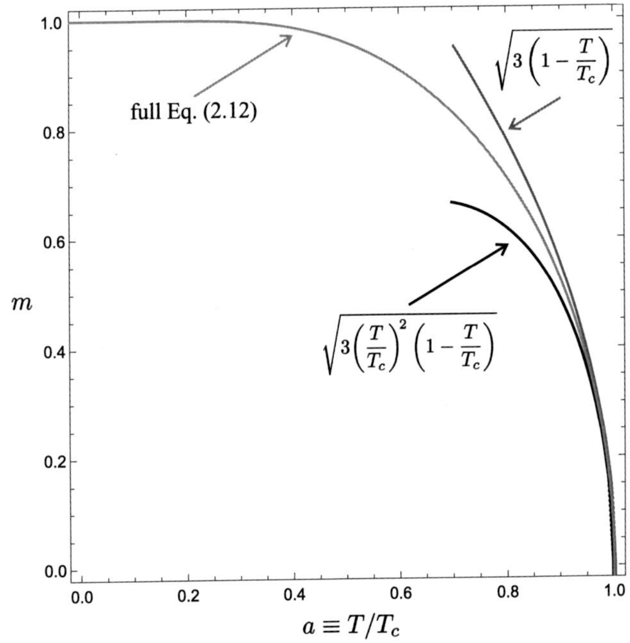

FIGURE 2.5 Temperature dependence of the order parameter m, together with its analytic estimate for $a \to 1$ (see Eq. (2.34)).

2.8 PROBLEM 2.2: A PHYSICAL ESTIMATE OF T_c-ORDERING ENERGY VS. ENTROPY

We have shown that, as $T \to T_c$ from below, the magnetic moment vanishes, $M \to 0$. This means that if the spin of the particle (or the atomic shell) is S, then, at $T \geq T_c$, all $(2S+1)$ orientations of the spin are equivalent when an applied magnetic field is absent, $B = 0$. Hence, in a system of N localized spins on atoms, the number of equivalent configurations is $\Omega = (2S+1)^N$. Estimate the value of T_c for the system of spins with the exchange integral $J > 0$ (i.e., for ferromagnetic interactions amongst the nearest neighbors).

Solution

The exchange energy is $E_{ex} = \langle H \rangle = -2Jz^2S^2N$, because

$$E_{ex} = \sum_{\langle ij \rangle} J_{ij} \langle S_i^z S_j^z \rangle = -JzNS^2. \tag{2.35}$$

This energy is counterbalanced by the thermal motion of spins of magnitude k_BTN for $T \geq T_c$. Therefore, we can assume that, at $T = T_c$, the exchange energy is compensated by the thermal energy, i.e.,

$$-JzS^2N = -TS_e \equiv -k_BT_cN \ln(2S+1), \tag{2.36}$$

where $S_e = k_B \ln \Omega$ is the system entropy.

In effect,

$$T_c = \frac{JzS^2}{k_B \ln(2S+1)}. \tag{2.37}$$

Note that T_c varies as J, and is proportional to the magnitude of the spin. The number of nearest neighbors here represents the lattice structure factor.

APPENDIX 2.A PRIMER IN STATISTICAL PHYSICS – A SUMMARY

In this Appendix, we assemble the most important concepts and formulas required to analyze the properties of the models considered in this book. Emphasis is put on the properties at temperature $T > 0$, encompassing in particular the phase-transition points. The understanding of the basic features of statistical physics is indispensable for understanding the theoretical material.

2.A.1 Principles of Thermodynamics (Thermostatics)

Historically, the fundamental concept in both thermodynamics and statistical mechanics is the *function of state*. The simplest example of such a function is the internal energy that represents a generalization of the concept of mechanical energy, since it also includes the effect of thermal (noise) energy pumped into the system from the environment. In the simplest case, the change of internal energy is represented by the two processes

$$dU = đL + đQ, \tag{2.A.1}$$

where the change of the internal energy dU is composed of the work ("labor") $đL$ done on the system, and $đQ$ is the concomitant/alternative change of the thermal noise pumped into the system. If the system performs work on the environment or releases thermal energy, then both $đL$ and $đQ$ are negative. The change dU is universal, irrespective of the process selected to change, whereas $đL$ and $đQ$ depend on the processes selected to carry the processes through. This is why we describe U as a function of state. Usually it is necessary to discuss the nature of $đL$ and $đQ$ because the derivatives of U (or related functions of state defined below) are directly measurable; U can be determined microscopically without going into detailed nature of the two terms composing it.

A question arises how a system can remain at equilibrium as it exchanges energy with the surroundings. In this regard, we claim that *a quasistatic (reversible) process* is composed of a sum of elementary steps following each other through a series of transient equilibria. This concept of a thermodynamic (strictly speaking, *thermostatic*) processes, composed of a series of incremental changes, will be implicitly assumed to occur throughout the book. This is one of the reasons why the principle (2.A.1) is expressed in differential (incremental) form. In the formulation (2.A.1), the energy is defined to be a constant (i.e., may contain other important terms). Even when the phase transition takes place in infinite incremental changes $đQ/dT$ of heat absorption/release at the critical temperature $T = T_c$, the quasistatic character of the process is assumed to be preserved. This assumption poses quite stringent conditions on the physical measurements in a regime very close to the critical point (in the so-called *critical regime*), but the basic experimental procedures required to achieve this regime are outside the purview of this book.

The choice of internal energy is an empirical generalization of the concept of mechanical energy, which encompasses the universal thermal chaos accompanying it. Its definition is not easy to grasp because it is composed in the simplest case of the work in the form $đL = -PdV$ (P is the pressure, and V is the volume), and therefore, depends on the system volume V. Additionally, the heat transfer is taken to be of the form $đQ = TdS$, where S is the system entropy. Effectively, $đQ$ depends on the two extensive variables, V and S. Throughout the book, we are mostly concerned with magnetic properties of materials. This introduces another work term for the magnetization of materials. In a magnetic system, the additional elementary work performed to magnetize the system further is $\boldsymbol{H}_a \cdot d\boldsymbol{M}$, where $\boldsymbol{H}_a$ is the applied magnetic field, and $\boldsymbol{M}$ is the system magnetization (total dipole magnetic moment). Therefore, instead defining the incremental change of the internal energy

$$dU = -p\,dV + T\,dS + \boldsymbol{H}_a \cdot d\boldsymbol{M}, \tag{2.A.2}$$

which depends on extensive (total) quantities, we use the relation $d(PV) = P\,dV + V\,dP$ (and analogously for the other terms) and introduce the new function of state called the Gibbs free energy $G \equiv U - TS + PV - \boldsymbol{H}_a \cdot d\boldsymbol{M}$, for which

$$dG = V\,dP - S\,dT - \boldsymbol{M} \cdot d\boldsymbol{H}_a, \tag{2.A.3}$$

so that $G = G(P, T, \boldsymbol{H}_a)$, i.e., is a function of only intensive (volume independent) variables. If additionally, we allow for an exchange of particles composing the system with the environment, then we have to add an additional term $+\mu\,dN$ to Eq. (2.A.3), where N is the number of particles, and μ is the so-called (electro)chemical potential of the system. In fact, $G = G(P, T, \boldsymbol{H}_a, N)$ represents the most general form of the function of state, which, when known, allows us to determine explicitly the principal system characteristics such as

$$S = -\left.\frac{\partial G}{\partial T}\right|_{P,\boldsymbol{H}_a,N}, \quad \mu = \left.\frac{\partial G}{\partial N}\right|_{P,\boldsymbol{H}_a,T}, \quad \boldsymbol{M} = -\left.\frac{\partial G}{\partial \boldsymbol{H}_a}\right|_{P,T,N}, \quad \text{etc.} \tag{2.A.4}$$

In practical applications, it is convenient to introduce also the Helmholtz free energy $F \equiv F(T, V, N) = U - TS$, for which we have

$$dF = -S\,dT - P\,dV + \boldsymbol{H}_a \cdot d\boldsymbol{M} + \mu\,dN. \tag{2.A.5}$$

The Helmholtz free energy is convenient for use in a situation with constant volume, whereas the Gibbs function is relevant for a study of the evolution of the system as a function of intensive variables $P, T, \boldsymbol{H}_a$, and N that are easy to control experimentally. In other words, we select the particular function of state reflecting a particular experimental situation.

2.A.2 Connection to Statistical Physics

The principles of the thermodynamics listed in the previous subsection provide only a framework for a description of real systems. Also, thermostatics does not provide any quantitative prescription for determining the functions of state or their derivatives. In essence, it allows us to determine the relations among the relevant quantities or functions. To determine explicitly the functions of state and other physical quantities, we require either the methods of statistical physics or postulating phenomenological forms of the function of state (e.g., the Landau theory), or even introducing phenomenologically an explicit equation of state (e.g., Clapeyron or van der Waals equation, etc.). To be precise, by *equation of state* we mean a functional relation between the fundamental thermodynamic variables (e.g., $P, V, T, \boldsymbol{M}$, and N) or any subset.

Statistical physics is based on either a microscopic Hamiltonian or on a Lagrangian. Here we select the Hamiltonian approach. For a canonical system, i.e., one with a constant number of a particles, we start from the partition function (the statistical sum) of the form

$$Z = \prod_n e^{-\beta E_n} \equiv \mathrm{Tr}\, e^{-\beta \hat{\mathcal{H}}}, \quad \beta \equiv 1/k_B T, \tag{2.A.6}$$

where E_n is the eigenenergy of the system state, characterized by the complete set of quantum numbers n, and the symbol "Tr" replaces the summation over eigenenergies $\{n\}$ if the Hamiltonian is not in diagonal form. The relation to statistical thermodynamics is established by making the following connection to the Helmholtz free energy:

$$F = -k_B T \ln Z. \tag{2.A.7}$$

As $T \to 0$, with E_0 being the ground state energy, we set $E_n = E_0 + (E_n - E_0)$, and with $(E_n - E_0) > 0$, we arrive at $F(T = 0) = E_0$, as should be the case. Note also that $F \equiv F(T, V)$.

When the particle number in the system fluctuates (i.e., when the system exchanges particles with the environment), we define the grand partition function as

$$\Xi = \sum_N \sum_n e^{-\beta(E_n - \mu N)} \equiv \mathrm{Tr}\, e^{-\beta(\hat{\mathcal{H}} - \mu \hat{N})}, \tag{2.A.8}$$

where the chemical potential μ is related to the fixed average value $\langle N \rangle$ of N in the following form:

$$\langle N \rangle = \sum_N \sum_n N e^{-\beta(E_n - \mu N)}. \tag{2.A.9}$$

Here $\langle N \rangle$ is identified as the physical number of particles in the system. The grand partition function is related to another function of state, the thermodynamic potential $\Omega \equiv \Omega(T, V, \mu)$, by

$$\Omega = -k_B T \ln \Xi. \tag{2.A.10}$$

Often, it is more convenient to work with the Gibbs function, to which Ω is related by:

$$G = \Omega + \mu \langle N \rangle. \tag{2.A.11}$$

All this analysis represents a summary of the essential ingredients of *statistical physics*.

The next task is to relate the above to particular physical quantities, such as the specific heat, the entropy, the magnetization, etc. This is discussed in the main text. Before summing up, let us note the following: When V and μ are constant (or, alternatively, P and N are constant), the functions of state F, G, and Ω are essentially identical functions of state, which all involve functions of T and $\boldsymbol{H}_a$ only. In other words, an appropriate function of state can be selected as the Helmholtz free energy $F(T, \boldsymbol{H}_a)$, a very convenient quantity for the description of magnetic systems with localized spins when magnetostriction effects can be ignored. This simple state cannot be achieved when we deal with itinerant fermions (e.g., with electrons in metals, metallic magnets, or superconductors). In that case it is usually more convenient to work with the grand canonical ensemble. If we can ignore the effect of changing pressure (or volume), then the most convenient function of state is the Gibbs free energy $F(T, \boldsymbol{H}_a, N)$. In the next subsection, we summarize the properties of quantum gases (i.e., itinerant fermions and bosons).

2.A.3 Elementary Example: Statistical Distribution Function for Quantum Multiparticle System: Fermions and Bosons

In this subsection, we demonstrate the utility of the grand canonical ensemble when applied to quantum gases. This is used in the variable particle-number approach, even though mass is strictly conserved for material particles (i.e., those with nonzero mass). Suppose that such a gas is represented by a diagonalized Hamiltonian[8] of the form

$$\mathcal{H} = \sum_{\boldsymbol{k}\sigma} \epsilon_{\boldsymbol{k}} n_{\boldsymbol{k},\sigma}, \tag{2.A.12}$$

where $\epsilon_{\boldsymbol{k}}$ is the energy of the particle with wavevector $\boldsymbol{k}$ (momentum $\hbar\boldsymbol{k}$), and $\hat{n}_{k\sigma}$ is the particle umber of the particles with wavevector $\boldsymbol{k}$ and spin quantum number $\sigma = \pm 1 \equiv \uparrow\downarrow$. To consider quantum systems, Bogoliubov (1946) introduced the generalized Hamiltonian $\mathcal{H} - \mu N$,

$$\mathcal{H} - \mu N = \sum_{\boldsymbol{k}\sigma} \epsilon_{\boldsymbol{k}} n_{\boldsymbol{k}\sigma} - \mu \sum_{\boldsymbol{k}\sigma} n_{\boldsymbol{k}\sigma} = \sum_{\boldsymbol{k}\sigma} (\epsilon_{\boldsymbol{k}} - \mu)\, n_{\boldsymbol{k}\sigma}. \tag{2.A.13}$$

In this expression, μ is still the chemical potential in Eq. (2.A.9), but now related directly to the dynamical part of the problem. Note that minimally μ is the

8. By stating that the Hamiltonian is in diagonal form we mean that its total energy is written in the form of Eq. (2.A.12), i.e., when the single-particle energy spectrum $\{\epsilon_{\boldsymbol{k}}\}$, as well as the occupancies $\{n_{\boldsymbol{k}\sigma}\}$ are known. This reduces the calculation of the system (internal) energy to a counting procedure detailed below.

reference energy for individual particle energies, and for electrons, reduces at $T = 0$ to the Fermi energy.

The main reason for introducing μ in Eq. (2.A.12) is to have it serve as a Lagrange multiplier in the variable-particle approach, by which we fix their actual number *a posteriori*. The variable particle number approach allows us to assume that each energy level $\epsilon_{\boldsymbol{k}}$ can be occupied to a maximal allowed value, even if we do not have enough particles available. This happens particularly in the case of fermions when the number M of available energy states exceeds the number of particles N. In those circumstances we write down the ground canonical partition function in the form

$$\Xi = \sum_{\{n_{\boldsymbol{k}\sigma}\}} \exp\left[-\beta \sum_{\boldsymbol{k}\sigma} (\epsilon_{\boldsymbol{k}} - \mu)\, n_{\boldsymbol{k}\sigma}\right], \tag{2.A.14}$$

where, by summing up over all possible sets of occupancies $\{n_{\boldsymbol{k}\sigma}\}$, we sum both over all available the system energies and over the particle numbers. Thus, for either fermions or bosons, with $n_{\boldsymbol{k}\sigma} = 0, 1$ or $n_{\boldsymbol{k}} = 0, 1, 2, \ldots$, respectively, we obtain

$$\Xi = \prod_{\boldsymbol{k}\sigma} \left[1 \pm e^{-\beta(\epsilon_{\boldsymbol{k}} - \mu)}\right]. \tag{2.A.15}$$

Then the total number of particles in the system is

$$\langle N \rangle = \sum_{\boldsymbol{k}\sigma} n_{\boldsymbol{k}\sigma}\, p\,\{\epsilon_{\boldsymbol{k}}, n_{\boldsymbol{k}\sigma}\} \equiv \sum_{\boldsymbol{k}\sigma} \langle n_{\boldsymbol{k}\sigma} \rangle, \tag{2.A.16}$$

where

$$p\,\{\epsilon_{\boldsymbol{k}}, n_{\boldsymbol{k}\sigma}\} = \frac{1}{\Xi}\, e^{-\beta(\epsilon_{\boldsymbol{k}} - \mu) n_{\boldsymbol{k}\sigma}}. \tag{2.A.17}$$

We can execute the summation in (2.A.17) to obtain the statistical distribution function in its explicit form, namely

$$\langle n_{\boldsymbol{k}\sigma} \rangle \equiv f_{\boldsymbol{k}\sigma} = \frac{1}{e^{\beta(\epsilon_{\boldsymbol{k}} - \mu)} \pm 1} \tag{2.A.18}$$

for fermions (+) and bosons (−), respectively.

The Gibbs function, the entropy, and other physical quantities can be determined in a straightforward manner, once we specify the distribution (2.A.17). Namely, the internal energy has the form

$$U(T) = \sum_{\boldsymbol{k}\sigma} f_{\boldsymbol{k}\sigma}\, \epsilon_{\boldsymbol{k}} \equiv 2 \sum_{\boldsymbol{k}} f_{\boldsymbol{k}}\, \epsilon_{\boldsymbol{k}}, \tag{2.A.19}$$

which simply counts the energies and the corresponding occupancies. The entropy has the form

$$S = -k_B \sum_{\boldsymbol{k}\sigma} \left[f_{\boldsymbol{k}\sigma} \ln f_{\boldsymbol{k}\sigma} \mp (1 \mp f_{\boldsymbol{k}\sigma}) \ln (1 \mp f_{\boldsymbol{k}\sigma}) \right], \tag{2.A.20}$$

where, in the case of spinless bosons, the spin index and the summation over σ should be dropped. In fact, we can define the Helmholtz free energy as $F = U - TS$ and thus obtain all relevant physical quantities. For example, the two ways of defining the specific heat at $\boldsymbol{H}_a = 0$ leads to the general form (not limited to gases only)

$$C_V = \left(\frac{\partial U}{\partial T} \right) \bigg|_V , \quad C_P = \left(\frac{dQ}{dT} \right) \bigg|_P = T \frac{\partial S}{\partial T} = -T \left(\frac{\partial^2 F}{\partial T^2} \right) \bigg|_P , \tag{2.A.21}$$

whereas the magnetic moment reads

$$M \equiv g \, \mu_B \, N \bar{S^z} = - \left(\frac{\partial F}{\partial \boldsymbol{H}_a} \right) \bigg|_{T,\,P} . \tag{2.A.22}$$

Other quantities are discussed in the main text. In the mean field approximation and for spin systems, it is more convenient to use the real-space language, as detailed in Chapters 2–4.

2.A.4 Outlook

Let us return to general approach outlined in Subsection 2.A.2. First, to construct a statistical–mechanical model, we need to know the system Hamiltonian, which is diagonalized subsequently, i.e., its eigenenergy spectrum $\{E_n\}$ must be obtained. The partition function (2.A.6) or (2.A.8) can then be determined, depending on the actual physical situation. This quantity is used in turn to determine thermodynamic functions of state (2.A.7) or (2.A.11) (together with condition (2.A.9)), respectively. At this stage, we can calculate measurable physical quantities, e.g., (2.A.21) or (2.A.22). Those quantities can be compared with the ones measured experimentally, provided that the model, as represented by Hamiltonian of the system, describes the main dynamic and thermodynamic processes taking place.

It is worth quoting two important general formulas for the specific heat and the magnetic susceptibility, namely

$$C = \left(\langle \hat{\mathcal{H}}^2 \rangle - \langle \hat{\mathcal{H}} \rangle^2 \right) / \, (k_B \, T) , \tag{2.A.23}$$

$$\chi = (g \mu_B)^2 \left[\left\langle (S^z)^2 \right\rangle - (\langle S^z \rangle)^2 \right] / \, (k_B \, T) . \tag{2.A.24}$$

Those global formulas show that the specific heat describes the system energy fluctuations, whereas the magnetic susceptibility characterizes the magnetic moment fluctuations. Both quantities are measured in units of $k_B T$ the thermal noise energy. These expressions characterize directly the fluctuations at thermodynamic equilibrium, even though the system is in thermal contact with its environment, which thus must also be at equilibrium at every stage of our measurements. Their divergence at the critical point is a "smoking gun" of a continuous phase transition. Carrying out measurements as required by Eqs. (2.A.23) and (2.A.24) near T_c determines the essential singularities appearing at T_c.

2.A.5 Final Note

The material presented in this Appendix 2.A is a brief summary of a detailed analysis in the books on statistical physics. An example of a careful, but not an easy, textbook is L.D. Landau and E.M. Lifshitz, *Statistical Physics*, 3rd Edition Part I (Pergamon Press, Oxford, 1980). The present Appendix contains a prescription how to use the concepts in practical applications and provides neither their conceptual basis nor the necessary detailed formal analysis.

APPENDIX 2.B EXPANSION OF EQ. (2.5) IN POWERS OF *M*

A. The expansion of $\cosh u$ to fourth-order terms reads

$$\cosh u = 1 + \frac{u^2}{2!} + \frac{u^4}{4!} + o(u^6). \tag{2.B.1}$$

The logarithmic expansion to fourth order has the form

$$\ln \cosh u = \ln\left[1 + \frac{u^2}{2}\left(1 + \frac{u^2}{12}\right)\right] = \frac{u^2}{2}\left(1 + \frac{u^2}{12}\right) - \frac{u^4}{8}$$
$$= \frac{u^2}{2} - \frac{u^4}{12} + o(u^6). \tag{2.B.2}$$

The expression for $\mathcal{F}$ now follows from the last expansion.

Comments

The approximation scheme used here, as applied to a two dimensional lattice with the use of $S = \pm\frac{1}{2}$, leads to a value of $k_B T = 1.44\,J$ as compared to the exact value of $k_B T = 2.27\,J$ as determined by Onsager (1944). This indicates that longer range interactions, neglected in the present approach, also contribute significantly to the total interaction energy. In the mean field approximation $k_B T = J$.

In the derivations of the present chapter it is tacitly assumed that the thermal degrees of freedom evolve gradually with increasing temperature until the ordering (exchange) energy is reached. This assertion needs to be revised: as $T \to T_c$ the critical fluctuations, ignored in the present chapter, completely dominate the physical characteristics. These fluctuations lead to critical (singular) phenomena treated in subsequent chapters. The mean field approach simply sets the energy scale and the overall character of events responding to the combined effects of an applied magnetic field and of temperature.

B. Consider the terms in Eq. (2.6) involving the factor 1/2, which reads

$$-Nk_BT\frac{(\beta JzMS)^2}{2}. \tag{2.B.3}$$

Then substitute $k_BT_c = JzS^2$ to find

$$-Nk_BT\frac{\left(\frac{T_c}{TS}\right)^2 M^2}{2}. \tag{2.B.4}$$

Next, use $k_BT \equiv k_BT_c\frac{T}{T_c} = JzS^2\frac{T}{T_c}$ and proceed similarly with the $\frac{u^2}{12}$ term. This establishes Eq. (2.7).

C. Define $t \equiv \frac{T-T_c}{T_c}$, where close to the critical temperature, $t \ll 1$. Then expression (2.7) for $\mathcal{F}_u$ may be recast by examining the multiplier of M^2 in Eq. (2.7):

$$1-\frac{T_c}{T} = 1-\frac{1}{1+t} \approx t = \frac{T}{T_c}-1. \tag{2.B.5}$$

Concerning the term in $\frac{u^4}{12}$, it is customary ignore the distinction between T and T_c. The above establishes Eq. (2.13).

D. Use Eq. (2.22), define $q \equiv (\frac{T_c}{T}-1)$, and set $M = M_0 + \delta M$ with $M_0^2 = -3t$, $t \equiv \frac{T}{T_c} - 1$. Then Eq. (2.22) reads

$$q(M_0+\delta M) = \frac{1}{3}(M_0+\delta M)^3 - \frac{\mu B}{k_BT} \approx \frac{1}{3}M_0^2(M_0+3\delta M) - \frac{\mu B}{k_BT}, \tag{2.B.6}$$

where in the cubic expansion, we discard terms in δM^2 and δM as being of higher order. Then

$$\begin{aligned} M_0+\delta M &= \frac{1}{3}\left(-3\frac{t}{q}\right)(M_0+3\delta M) - \frac{\mu B}{k_BTq} \\ &= (1+t)(M_0+3\delta M) - \frac{\mu B}{k_BTq}. \end{aligned} \tag{2.B.7}$$

We neglect t since it multiplies a term of first order. Equation (2.23) now follows.

Chapter 3

General Mean Field Approach

In starting off this chapter, we ask whether the findings of Chapter 2 are peculiar to the Ising model, or whether they are more generally valid, and, in particular, whether they apply to cases other than lattice configurations. We find that this is the case, which may come as a surprise, since that model fixates on spin interactions between nearest neighbors and with an externally applied magnetic field. It turns out that such an approach is representative of a whole class of models. It is therefore instructive to examine the mean field approach to critical phenomena in a broader context, again starting from a model with magnetic exchange interactions, this time in spin-rotationally invariant form in spin space.

3.1 THE HEISENBERG MODEL AND THE MEAN FIELD APPROXIMATION

We start from the Heisenberg Hamiltonian expressing the spin–spin interaction in a spin-rotationally invariant form, i.e.,

$$\hat{\mathcal{H}} = -J \sum_{i \neq j} \hat{\boldsymbol{S}}_i \cdot \hat{\boldsymbol{S}}_j - g\mu_B \boldsymbol{B} \cdot \sum_i \hat{\boldsymbol{S}}_i. \tag{3.1}$$

The interaction is assumed again as taking place between nearest neighbors only; thus, the summation $\sum_{i \neq j}$ is confined to nearest-neighbor pairs $\langle ij \rangle$. The spin operator located at the site i is $\hat{\boldsymbol{S}}_i \equiv \left(\hat{S}_i^x, \hat{S}_i^y, \hat{S}_i^z \right)$.

Within the above framework, we select a representative spin $\boldsymbol{S}_i$ that is then subjected to an externally applied magnetic field $\boldsymbol{B}$ and to the effective field generated by all other spins in the lattice. A fundamental simplification is introduced by replacing the individual spins by their average expectation value $\langle \hat{\boldsymbol{S}}_i \rangle$, an approach that comes under the heading of *mean field theory*. Also, given a single value of J, the spin interactions are confined to equidistant nearest-neighbor pairs, so that the total interaction energy is modeled by assigning the representative spin $\boldsymbol{S}_i$ the energy $E_i = -\hat{\boldsymbol{S}}_i \cdot J \sum_{j(i)} \langle \hat{\boldsymbol{S}}_j \rangle - g\mu_B \boldsymbol{B} \cdot \hat{\boldsymbol{S}}_i \equiv -g\mu_B \hat{\boldsymbol{S}}_i \left(\frac{J}{g\mu_B} \sum_{j(i)} \langle \hat{\boldsymbol{S}}_j \rangle + B \right) \equiv -g\mu_B \hat{\boldsymbol{S}}_i \cdot \boldsymbol{B}_{\mathrm{eff},i}$. Assuming that the *effective field is spatially homogeneous*, i.e., $\boldsymbol{B}_{\mathrm{eff},i} \equiv \boldsymbol{B}_{\mathrm{eff}}$,

A Primer to the Theory of Critical Phenomena. http://dx.doi.org/10.1016/B978-0-12-804685-2.00003-6

and selecting the spin quantization axis $e_z \parallel \boldsymbol{B}_{\text{eff}}$, the scalar product $\hat{\boldsymbol{S}}_i \cdot \boldsymbol{B}_{\text{eff}}$ can be replaced by $\hat{S}_i^z \cdot B_{\text{eff}}$, with $S_i^z = \pm\frac{1}{2}$. We then encounter two energy levels, $E_i^+ = -\frac{1}{2}\left[J\sum_j \langle S_j^z\rangle + g\mu_B B\right] = -\frac{1}{2}\left[JzM + g\mu_B B\right]$ and $E_i^- = \frac{1}{2}\left[J\sum_j \langle S_j^z\rangle + g\mu_B B\right] = \frac{1}{2}\left[JzM + g\mu_B B\right]$. Here the sum $\sum_{j(i)} \langle S_j^z\rangle$ involves the z nearest neighbors to spin $\hat{\boldsymbol{S}}_i$; as in Chapter 2, we have also set $\langle S_j\rangle \equiv M$, the "magnetization." We also reintroduce the effective magnetic field as $B_{\text{eff}} \equiv JzM + g\mu_B B$ (in energy units). The corresponding Hamiltonian then has the form

$$\mathcal{H} = -B_{\text{eff}} \sum_i \hat{S}_i^z, \tag{3.2}$$

and the corresponding integral energy $U \equiv \langle \hat{H}\rangle$ is

$$U = \frac{1}{2}\left(JzM^2 + g\mu_B MB\right) N, \tag{3.3}$$

where N is the number of spins. We see that the form (3.2) is now identical to that for the Ising spins with $S_i \equiv S_i^z$. Hence, the magnetization $M \equiv \langle S_i^z\rangle$ is determined from the same equation as for the Ising model within spin $S = 1/2$, namely

$$M = \frac{1}{2}\tanh\left(\frac{B_{\text{eff}}}{2k_B T}\right) = \frac{1}{2}\tanh\left(\frac{JzM + g\mu_B B}{2k_B T}\right). \tag{3.4}$$

Hence, at the mean field level at least, the Heisenberg and Ising models provide the same statistical thermodynamics for the spins. This is so because the angular fluctuations of the magnetic moments $\langle \hat{\boldsymbol{S}}_i\rangle$ are ignored (see the next chapters).

Important note

Note that the pairwise interaction in Eq. (3.1) has been replaced by the effective interaction of all nearest-neighboring spins $J(i)$ with a given spin located at the site i. Such a replacement of the dynamic variables $\{\hat{\boldsymbol{S}}_j\}$ by an effective field $\boldsymbol{B}_{\text{eff}}$, and thus the Hamiltonian (3.2), breaks the rotational symmetry in the spin space of the original Hamiltonian (3.1). This procedure amounts to introducing broken symmetry states, as discussed in the next chapter.

3.2 CRITICAL PHENOMENA IN THE MEAN FIELD APPROXIMATION: A FURTHER ELABORATION

We are now ready to investigate physical properties of the system close to the critical point. We note that the subsequent analysis of magnetic properties involves the magnetization M rather than the scaled equilibrium magnetization $\frac{M}{S} = 2M$. Hence we eliminate the factor 2 in the denominator of Eq. (3.4). We also set $\mu \equiv g\mu_B$ and introduce the reduced reciprocal temperature $\tau \equiv \frac{T_c}{T}$ to write

$$M = \tanh\left(\frac{\mu B}{k_B T} + M\tau\right) \equiv \frac{\tanh\left(\frac{\mu B}{k_B T}\right) + \tanh(M\tau)}{1 + \tanh\left(\frac{\mu B}{k_B T}\right) \cdot \tanh(M\tau)}, \tag{3.5}$$

where, as shown in Eq. (2.11) with $S = 1$, $k_B T_c = Jz$, and $\tau \equiv T_c/T$. The above relation may readily be recast in the form

$$\tanh\left(\frac{\mu B}{k_B T}\right) = \frac{M - \tanh(M\tau)}{1 - M\tanh(M\tau)}. \tag{3.6}$$

Now go to the limit of small B and set $\tanh(M\tau) \approx M\tau - \frac{(M\tau)^3}{3}$. The expansion is straightforward and leads to

$$\frac{\mu B}{k_B T} = M(1-\tau) + M^3\tau\left(1 - \tau + \frac{\tau^2}{3}\right). \tag{3.7}$$

We thus reach the Landau equation for the order parameter M in the presence of applied magnetic field. This sets the stage for specifying the critical exponents.

(i) Consider first the case $B = 0$:
Set $\tau = 1 - \delta$, $\delta \ll 1$; also, $1 - \frac{1}{\tau} = 1 - \frac{T}{T_c} = 1 - \frac{1}{1-\delta} \approx \delta$. The resulting expansions of Eq. (3.7) are straightforward; on discarding all terms beyond δ the physically relevant solution is given by

$$M \approx \sqrt{\frac{3(T_c - T)}{T_c}}, \quad T < T_c. \tag{3.8}$$

As before, this variation of the spontaneous magnetization with temperature near criticality is conventionally written out as[1] $M \sim (T_c - T)^\beta$; thus, in the mean field approximation, $\beta = \frac{1}{2}$.

(ii) Next, set $\tau = 1$, equivalent to being exactly at the critical temperature, and solve Eq. (3.7) for

$$\frac{\mu B}{k_B T} = \frac{M^3}{3}, \quad T = T_c. \tag{3.9}$$

This leads to the conclusion that $M \sim B^{\frac{1}{3}}$. It is customary to express the general relation as $M \sim B^{\frac{1}{\delta}}$, leading to the mean field assignment $\delta = 3$.

---

1. Here we run into notational problems. The critical exponent governing this relationship is almost universally assigned the symbol β, not to be confused with $\frac{1}{k_B T}$, which we have introduced earlier.

(iii) In the limit of vanishing but nonzero B, the isothermal magnetic susceptibility per spin χ_T is found via the total magnetization $N\mu M$:

$$\chi_T = \left(\frac{\partial N\mu M}{\partial B}\right)_T. \tag{3.10}$$

The simplest way to determine χ_T is to multiply Eq. (3.7) by $N\mu B$ and then differentiate with respect to B. We obtain

$$\frac{N\mu^2}{k_B T} = \chi_T(1-\tau) + 3M^2\chi_T\left(\tau - \tau^2 + \frac{\tau^3}{3}\right). \tag{3.11}$$

There are now two cases to consider:

(a) $T > T_c$, where $M = 0$, so that

$$\chi_T \approx \frac{1}{k_B}\frac{N\mu^2}{T - T_c}. \tag{3.12}$$

This may be compared to the empirically specified divergence $\chi_T \sim (T - T_c)^{-\gamma}$. We conclude that $\gamma = 1$ in the mean field approximation.

(b) $T < T_c$: In first approximation, set $\tau = 1$ in the second term of Eq. (3.11), which is appropriate since M is of first order of smallness. Then replace M^2 by Eq. (3.8) and solve for

$$\chi_T \approx \frac{1}{2k_B}\frac{N\mu^2}{T_c - T}. \tag{3.13}$$

Notice that again $\gamma = 1$ and that the two versions differ in sign and by a factor of two. Relation (3.13) is only half of Eq. (3.12) for the same value of $|T - T_c|$ since, for $T < T_c$, the spins are partially frozen in the z direction, an obvious sign of a spontaneously broken symmetry state $(M > 0)$.

(iv) To determine the magnetic contribution to the heat capacity C at constant volume and close to the critical point, we follow the arguments adduced in Chapter 2: C exhibits a discontinuity at $T = T_c$.

As is obvious, we encounter here precisely the same relations near criticality as for the Ising problem, which is not surprising in light of the similarity of effective field expression that appears in both cases when the magnetic order is spatially homogeneous.

Correlation function

Here we have our first formal look at correlation functions, whose importance will be discussed in Chapter 6 and demonstrated in following chapters.

We begin by considering an Ising system subjected to a position-dependent applied magnetic field B_k whose Hamiltonian has the form ($\hat{\mathcal{H}}_0 \equiv -J\sum_{\langle ij\rangle} S_i S_j$)

$$\hat{\mathcal{H}} = \hat{\mathcal{H}}_0 - \mu \sum_i S_i B_i \tag{3.14}$$

and for which the corresponding partition function is given by

$$\mathcal{Z}[B_i] = \sum_{\{S_i\}} \exp\left[-\beta \left(\hat{\mathcal{H}}_0 - \mu \sum_i S_i B_i \right) \right]. \tag{3.15}$$

Now execute the following sequence of operations:

$$\begin{aligned} \frac{1}{\beta\mu\mathcal{Z}}\left(\frac{\partial \mathcal{Z}}{\partial B_i}\right) &= \frac{1}{\beta\mu\mathcal{Z}}\frac{\partial}{\partial B_i}\sum_{S_k} e^{-\beta\left(\hat{\mathcal{H}}_0 - \mu\sum_k S_k B_k\right)} \\ &= \frac{1}{\mathcal{Z}}\sum_{S_k} S_i e^{-\beta\left(\hat{\mathcal{H}}_0 - \mu\sum_k S_k B_k\right)} \equiv \langle S_i \rangle. \end{aligned} \tag{3.16}$$

Thus, the above provides a method for finding the expectation value for a representative spin, as given by

$$\langle S_i \rangle = \frac{1}{\beta\mu\mathcal{Z}}\frac{\partial \mathcal{Z}}{\partial B_i} = \frac{1}{\beta\mu}\frac{\partial \ln \mathcal{Z}}{\partial B_i}. \tag{3.17}$$

Incidentally, this same method can be invoked even in cases where no external field is present, as explained in Appendix 3.B. By analogy the expectation value for $\langle S_i\, S_j \rangle$ involves the standard expression

$$\langle S_i S_j \rangle = \frac{1}{\mathcal{Z}}\sum_{S_k} S_i S_j \exp\left[-\beta \left(\hat{\mathcal{H}}_0 - \mu \sum_k S_k B_k \right) \right]. \tag{3.18}$$

This relation is obtained by executing a second differentiation of Eq. (3.17) analogous to Eq. (3.16), as follows:

$$\langle S_i S_j \rangle = \frac{1}{(\beta\mu)^2\mathcal{Z}}\frac{\partial^2 \mathcal{Z}}{\partial B_i \partial B_j} \equiv G(i, j), \tag{3.19}$$

which thus defines the *correlation function* $G(i, j)$.

Note further that Eq. (3.18) includes all the terms for which $j = i$; thus, to concentrate on the correlations involving spins on different sites, we define and construct a *connected correlation function* in the following manner:

$$G_c(i, j) \equiv \frac{1}{(\beta\mu)^2\mathcal{Z}}\frac{\partial^2 \mathcal{Z}}{\partial B_i \partial B_j} - \left(\frac{1}{\beta\mu\mathcal{Z}}\frac{\partial \mathcal{Z}}{\partial B_i}\right)\left(\frac{1}{\beta\mu\mathcal{Z}}\frac{\partial \mathcal{Z}}{\partial B_j}\right). \tag{3.20}$$

Note an alternative representation of Eq. (3.20),

$$G_c(i, j) = \frac{1}{(\beta\mu)^2} \frac{\partial^2 \ln \mathcal{Z}}{\partial B_i \partial B_j}, \tag{3.21}$$

which is verified by carrying out the indicated mathematical operations. The function $G_c(i, j)$ defined by Eq. (3.20) may also be represented by

$$G_c(i, j) = \langle S_i S_j \rangle - \langle S_i \rangle \langle S_j \rangle. \tag{3.22}$$

In other words, the entity $G_c(i, j)$ specifies the expectation value for two spins on sites i and j interacting with each other, relative to their individual expectation values on those sites. Some authors in discussing correlations refer to formulation (3.22) as correlation functions.

A very important additional relation is found by differentiating Eq. (3.17) using Eq. (3.21):

$$\frac{\partial \langle S_i \rangle}{\partial B_j} = \frac{1}{\beta\mu} \frac{\partial^2 \ln \mathcal{Z}}{\partial B_i \partial B_j} = \beta\mu G_c(i, j). \tag{3.23}$$

Now introduce the total magnetic moment $\mathcal{M} = NM$ and total spin expectation value $\langle \mathcal{S} \rangle$ as

$$\mathcal{M} = g\mu_B \sum_j \langle S_j \rangle \equiv \mu \langle \mathcal{S} \rangle \tag{3.24}$$

and then differentiate:

$$\frac{\partial \mathcal{M}}{\partial B_j} = \mu \sum_j \frac{\partial \langle S_j \rangle}{\partial B_j} = \beta\mu^2 \sum_i G_c(i, j). \tag{3.25}$$

For the particular case of a uniform field B, we find that

$$\frac{\partial \mathcal{M}}{\partial B} = \sum_j \frac{\partial \mathcal{M}}{\partial B_j} \frac{\partial B_j}{\partial B} = \beta\mu^2 \sum_i \sum_j G_c(i, j), \tag{3.26}$$

which yields the isothermal magnetic susceptibility as

$$\begin{aligned} \chi_T &= \beta\mu^2 \sum_i \sum_j G_c(i, j) \equiv \beta\mu^2 \sum_i \sum_j G_c(r_i - r_j) \\ &= \beta\mu^2 \left[\sum_i \sum_j \langle S_i S_j \rangle - \left(\sum_i \langle S_i \rangle \right)^2 \right] = \beta\mu^2 \left(\langle \mathcal{S}^2 \rangle - \langle \mathcal{S} \rangle^2 \right). \end{aligned} \tag{3.27}$$

The expressions on the right obtain because the magnetic field is uniform.

The above relation relates $G_c(i, j)$ and the indicated fluctuations of the total spin (its standard deviation) to the experimentally observable isothermal magnetic susceptibility χ_T. Equation (3.27) is a representation of the *static susceptibility sum rule*, which links the thermodynamic observable χ_T to fluctuations in terms of the corresponding correlation function.

3.3 VAN DER WAALS EQUATION OF STATE AND CRITICALITY*

We next consider the case of the phenomenological van der Waals fluid, which is a departure from the lattice model, with spin–spin interactions between the constituents. The van der Waals equation of state is, to quote N. Goldenfeld[2] in a different context, the *drosophila* of the theory of fluids. On a molar basis, it has the standard form ($\tilde{V}$ is the molar volume)

$$P = \frac{RT}{\tilde{V} - b} - \frac{a}{\tilde{V}^2}, \tag{3.28}$$

for which the variation of pressure P with volume is sketched for a series of temperatures in Fig. 3.1. Note that, for $a = b = 0$, we obtain the Clapeyron equation for a classical ideal gas. The parameters a and b must be determined experimentally; $R = 8.31$ J/mol K is the gas constant. The dashed dome-like structure envelops the two-phase portion of the diagram; this is an exclusion region that is not encountered under equilibrium conditions. Liquid and vapor linked by the common tie-line are at equilibrium. Beyond its confines, we encounter the single phase domain of the P–$\tilde{V}$ diagram as marked. The curves represent the isotherms. Of particular interest is the curve that grazes the top of the dome with zero slope. This inflection point corresponds to the isotherm at the critical temperature T_c, where its first and second derivatives vanish. On imposing the requirements

$$\left(\frac{\partial P}{\partial \tilde{V}}\right)_{T_c} = \left(\frac{\partial^2 P}{\partial \tilde{V}^2}\right)_{T_c} = 0, \tag{3.29}$$

we find that

$$-\frac{2a}{\tilde{V}_c^3} + \frac{RT_c}{(\tilde{V}_c - b)^2} = 0, \tag{3.30}$$

$$-\frac{3a}{\tilde{V}_c^4} + \frac{RT_c}{(\tilde{V}_c - b)^3} = 0. \tag{3.31}$$

---

2. Nigel Goldenfeld, *Lectures on Phase Transitions and the Renormalization Group* (Perseus Books, Reading, MA, 1992).

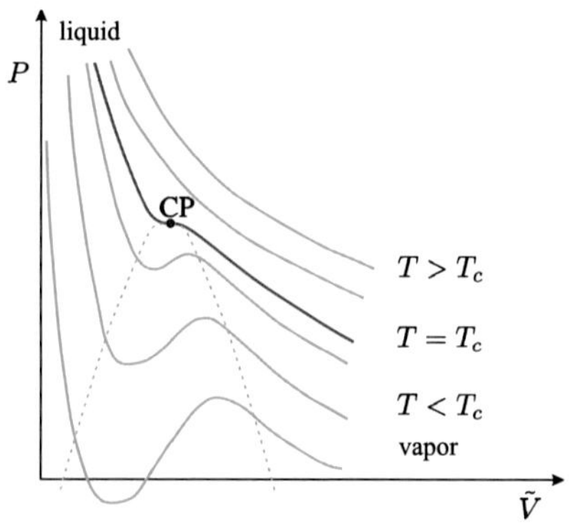

FIGURE 3.1 Pictorial representation of van der Waals isotherms on a pressure vs. molar volume diagram. The critical point is identified by CP. The region enclosed by the dashed curve is thermodynamically unstable. Note the shape of the isotherms above, at, and below the critical temperature T_c.

Division of (3.31) by (3.30) leads to the relation

$$\tilde{V}_c = 3b. \tag{3.32}$$

Inserting this into (3.30), we obtain

$$T_c = \frac{8a}{27bR}. \tag{3.33}$$

Substitution of (3.32) and (3.33) into the van der Waals equation of state leads to

$$P_c = \frac{a}{27b^2}. \tag{3.34}$$

Then the ratio

$$\frac{RT_c}{P_c}\tilde{V}_c = \frac{8}{3} \tag{3.35}$$

is a universal constant, whereas the individual critical constants do depend on the parameters a and b. Moreover, it is not difficult to show that, on introducing the scaled or reduced variables $\pi \equiv \frac{P}{P_c}$, $v \equiv \frac{\tilde{V}}{\tilde{V}_c}$, and $\tilde{t} \equiv \frac{T}{T_c}$, the van der Waals equation of state assumes the universal form

$$\left(\pi + \frac{3}{v^2}\right)(3v - 1) = 8\tilde{t}, \tag{3.36}$$

which is a very elegant expression since it involves solely the critical constants of fluids as the relevant parameters. Equation (3.36) is an example of the so-called *law of corresponding states*; if applicable, the scaled data for all fluids

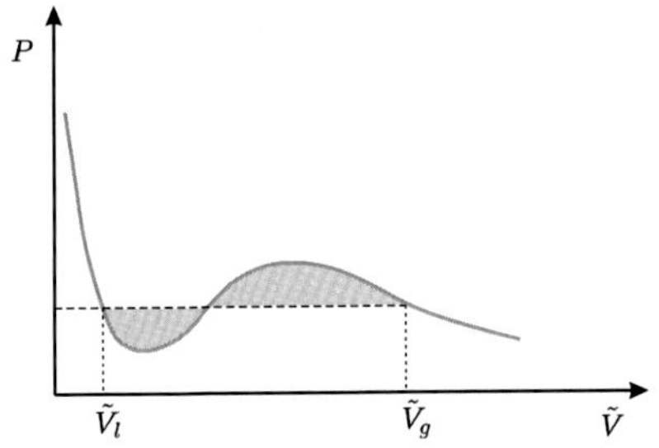

FIGURE 3.2 Sketch showing the Maxwell equal-area construction for an isotherm with $T < T_c$.

should fall on a single curve that illustrates the universality of some physical properties of a material. The extent to which this is the case, along with its shortcomings, was already illustrated in Chap. 1 for one class of materials. Other commonly used equations of state give rise to similar results but suffer from similar defects. We will rectify this problem once we proceed beyond mean field theory.

Now consider small deviations $t = \tilde{t} - 1 = \frac{T-T_c}{T_c}$ and $\tilde{v} \equiv v - 1 = \frac{\tilde{V}-\tilde{V}_c}{\tilde{V}_c}$ close to criticality where $\pi = \tilde{t} = v = 1$. In terms of these parameters the reduced equation of state becomes

$$\pi = \frac{8\,(1+t)}{3\,(1+\tilde{v})-1} - \frac{3}{(1+\tilde{v})^2}, \tag{3.37}$$

which, near the critical point, may be expanded as

$$\pi = 1 + 4t - 6\tilde{v}t - 3\frac{\tilde{v}^3}{2} + 9\tilde{v}^2 t + 27\frac{\tilde{v}^3 t}{2}. \tag{3.38}$$

We may discard the last two terms. The scaled coexistence volumes $\tilde{v}_l(P)$ and $\tilde{v}_g(P)$ for the liquid and vapor phases are in equilibrium at any fixed $T < T_c$.

Critical exponents for the van der Waals fluid

(i) We now relate the properties of the liquid and vapor phases of a van der Waals fluid: with reference to Fig. 3.2 (which is plotted in a highly schematic manner for purely illustrative purposes), we apply Maxwell's equal-area rule to require that, for fixed $\tilde{t} < 1$,

$$\int_{P_l}^{P_g} V\,dP = 0 = V_c \int_{P_l}^{P_g} \left(\frac{V}{V_c} - 1\right) dP \equiv \mathrm{I}. \tag{3.39}$$

The enlarged integral I vanishes because $\int_{P_l}^{P_g} dP = 0$ over the horizontal path indicated in the figure. But by Eq. (3.38) we find that $dP = -P_c\left(6t + 9\frac{\tilde{v}^2}{2}\right) d\tilde{v}$, whence I may be reformulated as

$$\int_{\tilde{v}_l}^{\tilde{v}_g} P_c \tilde{v}\left(6t + 9\frac{\tilde{v}^2}{2}\right) d\tilde{v} = 0. \tag{3.40}$$

This equation must be satisfied for all fixed t for which the approximation (3.38) holds. On integrating we are led to the requirement

$$3t\left(\tilde{v}_g^2 - \tilde{v}_l^2\right) + \frac{9}{8}\left(\tilde{v}_g^4 - \tilde{v}_l^4\right) = 0, \tag{3.41}$$

which is met by setting $\tilde{v}_g = \pm\tilde{v}_l$. We discard the positive root, which leads to an identity. On applying the condition $\tilde{v}_g = -\tilde{v}_l$ to Eq. (3.38) we obtain

$$\pi_g = 1 + 4t - 6t\tilde{v}_g - \frac{3\tilde{v}_g^3}{2}; \quad \pi_l = 1 + 4t + 6t\tilde{v}_g + \frac{3\tilde{v}_g^3}{2}. \tag{3.42}$$

Equating the two expressions and then solving for $\tilde{v}_g$, we find that, in the two-phase region $T < T_c$,

$$\tilde{v}_g = 2(-t)^{\frac{1}{2}} = 2\left[\frac{T_c - T}{T_c}\right]^{\frac{1}{2}} = -\tilde{v}_l \quad (T < T_c;\ t < 0), \tag{3.43}$$

showing that the scaled molar volume should change as the square root of small deviations of T below its critical value. This is to be compared to the empirical relation as written out in conventional form:

$$\tilde{v} \sim [T_c - T]^{\beta} \ \text{ with } \beta = \frac{1}{2}. \tag{3.44}$$

As before, β is a *critical exponent*, relating to the volume changes with temperature as the critical point is approached from below. As is readily shown by an examination of Fig. 3.2, an alternative formulation involving corresponding densities ρ reads

$$\rho_l - \rho_g \sim [T_c - T]^{\frac{1}{2}}. \tag{3.45}$$

These relations differ from the experimentally observed values that are close to $\beta = 0.328$.

(ii) Along the same lines, we can determine the pressure–volume relationship at the critical temperature by setting $T = T_c$ and $t = 0$ in Eq. (3.38). Conventionally, this relationship is expressed in the form

$$\pi - 1 = \frac{P - P_c}{P_c} = -\frac{3}{2}\left[\frac{V - V_c}{V_c}\right]^{\delta} \ \text{ with } \delta = 3, \tag{3.46}$$

which also disagrees with the observation that the experimentally observed critical exponent for the scaled pressure vs. scaled volume lies in the vicinity of $\delta = 4.8$ for binary fluids.

(iii) We now consider the heat capacity of van der Waals fluids at their respective critical points. One approach is to use the well-established thermodynamic relation between the molar heat capacity at constant pressure $\tilde{C}_P$ and the molar heat capacity $\tilde{C}_V$ at constant volume: Standard thermodynamic analysis shows that, in the single-phase region with $T > T_c$, this difference is specified by

$$\tilde{C}_P - \tilde{C}_V = -T \frac{\left(\frac{\partial \tilde{V}}{\partial T}\right)_P^2}{\left(\frac{\partial \tilde{V}}{\partial \tilde{P}}\right)_T}. \tag{3.47}$$

On inserting the requisite relations derived from the equation of state, as shown in Appendix 3.A, we find that very close to the critical point

$$\tilde{C}_P = \tilde{C}_V + \frac{RT}{T - T_c}, \tag{3.48}$$

where R is the gas constant. This is to be compared with the standard formulation $C_P \sim (T - T_C)^{-\alpha}$, whence $\alpha = 1$. Equation (3.48) applies only in the range $T > T_c$; at lower temperatures the two-phase nature of the system gets in the way of a simple analysis. For a majority of fluids, the experimental values of α fall in the range of 0.11.

(iv) The fourth measurement pertains to the isothermal compressibility κ_T. In the range $T < T_c$, we take the derivative of Eq. (3.46) with respect to P:

$$\frac{1}{P_c} = -\frac{3}{2V_c^3} 3(V - V_c)^2 \left(\frac{\partial V}{\partial P}\right) = \frac{9}{2}\left(\frac{V - V_c}{V_c}\right)^2 \kappa_T. \tag{3.49}$$

Now introduce Eq. (3.43) and solve for

$$\kappa_T = \frac{1}{V}\left(\frac{\partial V}{\partial T}\right)_N = \frac{T_c}{18 P_c (T_c - T)}. \tag{3.50}$$

In the range above T_c, we revert to Appendix 3.A to find that ($\tilde{V} \equiv \frac{V}{N}$)

$$\left(\frac{\partial \tilde{V}}{\partial P}\right)_T = \left[\frac{2a}{\tilde{V}^2} - \frac{RT}{(\tilde{V} - b)^2}\right]^{-1} = -\frac{\kappa_T}{\tilde{V}}. \tag{3.51}$$

Now replace $\tilde{V}$ by $\tilde{V}_c$ and replace a and b, as specified in Appendix 3.A. After considerable simplification this leads to

$$\kappa_T = \frac{4\tilde{V}_c}{9R(T - T_c)}. \tag{3.52}$$

We may compare this equation with Eq. (3.50) by introducing in Eq. (3.52) the relation $\frac{P_c \tilde{V}_c}{RT_c} = \frac{3}{8}$, which applies to the van der Waals equation of state. The result is

$$\kappa_T = \frac{4}{27} \frac{T_c}{P_c(T_c - T)}. \tag{3.53}$$

Aside from the sign change the two expressions differ by a factor of 3 in the denominator.

In both cases the isothermal compressibility near the critical point diverges as $\kappa_T \sim |T - T_c|^{-1}$. This may be compared to the standard notation $\kappa_T \sim |T_c - T|^{-\gamma}$, whence $\gamma = 1$ in mean field theory. The experimental values lie in the range $\gamma = 1.23$ to 1.25.

These findings show that the van der Waals equation of state for a liquid model at the critical point leads to critical exponents in complete agreement with the corresponding mean-field values for the ferromagnet cited earlier and for the Ising lattice in Chapter 2. Thus, the microscopic properties of the system are irrelevant in the vicinity of their respective critical points. This feature turns out to be universally true and hence furnishes a guidepost for our later theoretical development. However, the predicted exponents deviate from experimental observations, leading us to a search for a better theoretical model for the system behavior as $T \to T_c$, i.e., in the vicinity of the critical (phase transition) point.

3.4 DENSITY FLUCTUATIONS AND COMPRESSIBILITY

We can enlarge on the above by considering density fluctuations. For this purpose, examine the following sequence of operations based on standard thermodynamic relations: the Gibbs free energy is specified by

$$dG = -SdT + VdP + \mu_c dN, \tag{3.54}$$

where μ_c is the chemical potential of the species. Equation (3.54) allows us to establish the following Maxwell relation (obtained taking second derivatives of

dG with respect to P and N in either order):

$$\left(\frac{\partial \mu_c}{\partial P}\right)_{T,N} = \left(\frac{\partial V}{\partial N}\right)_{T,P} = \frac{1}{\rho}, \tag{3.55}$$

where ρ is the number density of the particles in the system.

We shall be interested in the following quantity, obtained by a succession of mathematical equalities:

$$\begin{aligned}\left(\frac{\partial \mu_c}{\partial N}\right)_{V,T} &= -\frac{\left(\frac{\partial V}{\partial N}\right)_{T,P}}{\left(\frac{\partial V}{\partial \mu_c}\right)_{T,N}} = -\frac{\frac{1}{\rho}}{\left(\frac{\partial V}{\partial P}\right)_{T,N}\left(\frac{\partial P}{\partial \mu_c}\right)_{T,N}} \\ &= \frac{1}{\rho\kappa_T V}\left(\frac{\partial \mu_c}{\partial P}\right)_{T,N} = \frac{1}{\rho^2\kappa_T V},\end{aligned} \tag{3.56}$$

where we introduced Eq. (3.55) and κ_T, as defined in Eq. (3.51).

We next examine the mean number of particles present in the fixed volume V. This is given by the relation

$$\langle N\rangle = k_B T\frac{\partial \ln \Xi}{\partial \mu_c}, \qquad \Xi \equiv \mathrm{Tr}\exp\left[-\beta(\hat{\mathcal{H}} - \mu_c N)\right], \tag{3.57}$$

where Ξ is the grand canonical partition function that takes account of density changes in the system. The "Tr" operation involves a summation over all relevant degrees of freedom.

In standard fashion the expectation value for N^2 is determined via

$$\langle N^2\rangle = \frac{\mathrm{Tr}N^2\exp[-\beta(\hat{H} - \mu_c N)]}{\mathrm{Tr}\exp[-\beta(\hat{H} - \mu_c N)]} = \frac{1}{\beta^2\Xi}\frac{\partial^2 \Xi}{\partial \mu_c^2} = \frac{1}{\beta^2}\frac{\partial^2 \ln \Xi}{\partial \mu_c^2} + \langle N\rangle^2. \tag{3.58}$$

Using Eqs. (3.55)–(3.57) with $\langle N\rangle = N$, we find that

$$\langle N^2\rangle - \langle N\rangle^2 = \frac{1}{\beta^2}\frac{\partial^2 \ln \Xi}{\partial \mu_c^2} = \frac{1}{\beta}\left(\frac{\partial N}{\partial \mu_c}\right)_{V,T} = k_B T\rho^2\kappa_T V. \tag{3.59}$$

This formulation shows how the fluctuation in particle numbers, $\langle N^2\rangle - \langle N\rangle^2$, is related to the isothermal compressibility, a quantity that can be measured.

We can also invoke the correlation function, which involves ρ in this case:

$$G(r - r') = \frac{1}{\rho^2}\left[\langle \rho(r)\rho(r')\rangle - \rho^2\right] \equiv \frac{1}{\rho^2}\langle(\rho(r) - \rho)(\rho(r') - \rho)\rangle, \tag{3.60}$$

which shows that the correlation function relates to fluctuations about the mean density $\langle \rho(r)\rangle = \rho$.

3.5 MEAN FIELD THEORY FOR BINARY MIXTURES*

We next present an outline showing how mean field theory can be applied to analyze the critical properties of a binary solution that separates into two immiscible components at low temperatures. Specifically, we consider a two-component mixture, labeled 1 and 2, that is essentially immiscible at ultralow temperatures. As the system is warmed up the extent of the immiscibility steadily decreases, until a critical temperature is reached where the last vestiges of the immiscibility disappear; above that point the solution forms a homogeneous single phase over the entire composition range.

Consistent with chemical usage, we specify the overall composition of the mixture in terms of n_1 moles of component 1 and n_2 moles of component 2, corresponding to mole fractions $x_i \equiv \frac{n_i}{(n_1+n_2)}$, $i = 1, 2$, with $0 \leq x_i \leq 1$ and $x_1 + x_2 = 1$. The Gibbs free energy G_m of a single phase that satisfies all the criteria of an ideal binary mixture is then spelled out in conventional fashion as

$$G_m = G_m^* + RT\left[n_1 \ln x_1 + n_2 \ln x_2\right], \tag{3.61}$$

where R is the gas constant, and G_m^* is a convenient reference value of no immediate interest. The logarithmic formula for the entropy pair is derived in Appendix 3.D. We introduce three minor changes: define the free energy of mixing as $\Delta G_m \equiv G_m - G_m^*$ and define the molar Gibbs free energy as $\tilde{G}_m = \frac{G_m}{(n_1+n_2)}$. Also, deviations from ideality are conventionally handled by replacing mole fractions x_i with *activities* $a_i = x_i \Gamma_i$, where Γ_i is termed an *activity coefficient*. The above relation thereby becomes

$$\begin{aligned} \frac{\Delta \tilde{G}_m}{RT} = \frac{\tilde{G}_m - \tilde{G}^*}{RT} &= x_1 \ln(x_1\Gamma_1) + x_2 \ln(x_2\Gamma_2) = \\ &= x_1 \ln x_1 + x_2 \ln x_2 + x_1 \ln \Gamma_1 + x_2 \ln \Gamma_2. \end{aligned} \tag{3.62}$$

Explicit relations for Γ_1 and Γ_2 are obtained by expanding $\ln \Gamma_1$ and $\ln \Gamma_2$ in a power series in x_1 and x_2 as follows:

$$\begin{aligned} \ln \Gamma_1 &= \ln a_1 - \ln x_1 = D_1 x_2 + B_1 x_2^2 + \cdots, \\ \ln \Gamma_2 &= \ln a_2 - \ln x_2 = D_2 x_1 + B_2 x_1^2 + \cdots. \end{aligned} \tag{3.63}$$

Observe that $\ln \Gamma_1$ is expanded as an ascending power series in x_2, and $\ln \Gamma_2$, in an ascending series in x_1. This recognizes that in dilute concentrations, where $x_2 \to 0$, component 1 exhibits ideal behavior $\Gamma_1 \to 1$; similarly, as $x_1 \to 0$, $\Gamma_2 \to 1$.

The coefficients on the right of (3.63) are not arbitrarily adjustable. To see this, set $i = 1$, take the differential form of Eq. (3.63), and multiply through by x_i to obtain

$$x_i d\ln a_i = x_i d\ln x_i - (D_i + 2B_i)x_i dx_i + 2B_i x_i^2 dx_i \quad (i = 1, 2). \tag{3.64}$$

From (3.64) form the sum $x_1 d\ln a_1 + x_2 d\ln a_2$, which must be forced to vanish at constant T and P in order to satisfy the Gibbs–Duhem relation, cited in the Appendix 3.C. Also, note that $x_1 d\ln x_1 + x_2 d\ln x_2 = dx_1 + dx_2 = 0$. Thus,

$$\begin{aligned} x_1 d\ln a_1 + x_2 d\ln a_2 = 0 = -(D_1 + 2B_1)\, x_1\, dx_1 - (D_2 + 2B_2)\, x_2\, dx_2 + \\ 2B_1\, x_1^2\, dx_1 + 2B_2\, x_2^2\, dx_2. \end{aligned} \tag{3.65}$$

Now set $x_2 = 1 - x_1$, cancel out the common multiplier dx_1, and rearrange the resulting terms in ascending powers of x_1, whereby

$$D_2 - [D_1 + D_2 + 2(B_1 - B_2)]\, x_1 + 2(B_1 - B_2)x_1^2 = 0. \tag{3.66}$$

The coefficients of $x_1^0 = 1,\ x_1^1 \equiv x_1$, and x_1^2 must vanish separately if (3.66) is to vanish for arbitrary x_1 in the interval $0 \le x_1 \le 1$. This requirement may be met by setting $D_1 = D_2 = 0$ and $B_1 = B_2 \equiv B$, which leads via Eq. (3.63) directly to the Margules (1895) equations

$$\ln\Gamma_1 = Bx_2^2 \quad \text{and} \quad \ln\Gamma_2 = Bx_1^2. \tag{3.67}$$

On introducing Eq. (3.67), Eq. (3.63) assumes the form $[x_1 = (1 - x_2),\ x_2 \equiv x]$,

$$\begin{aligned} \frac{\Delta\tilde{G}_m}{RT} &= (1-x)\ln(1-x) + x\ln x + Bx(1-x) \\ &= (1-x)\ln(1-x) + x\ln x + \frac{w}{RT}x(1-x). \end{aligned} \tag{3.68}$$

Here we have introduced a new parameter $w \equiv RTB > 0$. This fundamental equation forms the bed rock of our subsequent discussion.

Pictorial representations of the Margules relations

To investigate the implications of the above derivation, consider Fig. 3.3, which plots $\frac{\Delta\tilde{G}}{RT}$ vs. x for various $\frac{RT}{w} \equiv B^{-1} > 0$. As long as $\frac{RT}{w} > 0.5$, we encounter a U-shaped curve, indicating that a uniform mixture of any intermediate composition $x < 1$ is more stable than a comparable combination of pure (unmixed) ingredients. However, a new feature emerges for $\frac{RT}{w} < 0.5$: the presence of a double minimum that envelops a prominent maximum centered on $x = \frac{1}{2}$. We have encountered such a situation several times before: let the mole fractions

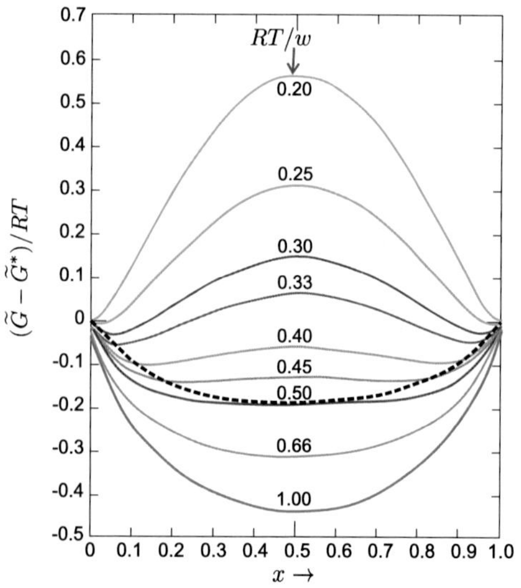

FIGURE 3.3 The variation of the molar Gibbs free energy of mixing with composition for various $\frac{RT}{w} > 0$. Dashed curve shows the locus of the double minima for $\frac{RT}{w} < 0.50$.

x' and $x'' = 1 - x'$ designate the locations of the two symmetrically disposed minima; then any mixture with composition $x' < x < x''$ is unstable. Correspondingly, in this range of $\frac{RT}{w}$ and composition, the binary mixture breaks into two immiscible components at mole fractions x' and x''; only homogeneous mixtures of composition outside this range are stable. The compositions x' and x'' of the immiscible constituents are determined as shown below.

The locus of the minima, shown as the dashed line in Fig. 3.3, is replotted in the $\frac{RT}{w}$ vs. x format in Fig. 3.4. Any composition within the dome is not encountered under equilibrium conditions. The homogeneity range in composition extends over the entire mole fraction range only when $\frac{RT}{w} > 0.5$.

We now proceed with a more quantitative analysis. The specific compositions x' and x'' of the heterogeneous phases are determined by the fact that x is a minimum at these two points: $\left(\frac{\partial \tilde{G}_m}{\partial x}\right)_{x''} = 0$. On imposing this requirement on Eq. (3.68) we arrive at

$$\ln\left[\frac{x''}{1-x''}\right] = B(2x''-1), \tag{3.69}$$

which is a transcendental equation that must be solved numerically for $x = x''$ once B is specified. As B is increased, x' and x'' move symmetrically away from the value $\frac{1}{2}$ toward 0 and 1, respectively.

Of immediate interest is the critical value B_c that separates the U-shaped curves from those that have their minima away from $x = \frac{1}{2}$. Accordingly, we

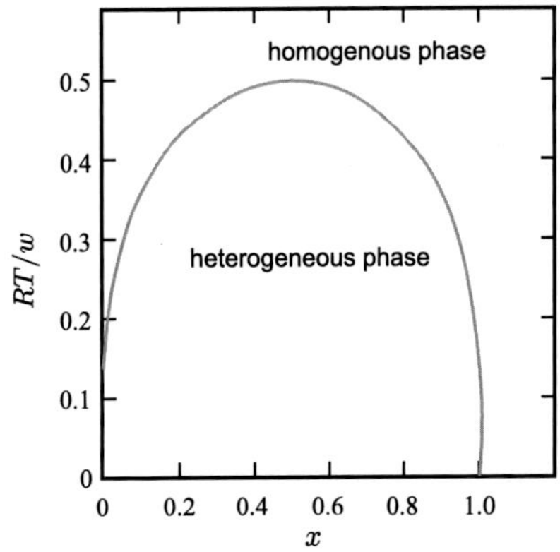

FIGURE 3.4 Plot showing the stability range of a binary mixture for the dependence of the scaled temperature vs. mole fraction of the second phase. The composition range within the dome is inaccessible at equilibrium.

impose two requirements on Eq. (3.68):

$$\frac{\partial^2 \Delta \tilde{G}_m}{\partial x^2} = 0 = \frac{1}{1 - x_c} + \frac{1}{x_c} - 2B_c \tag{3.70}$$

and

$$\frac{\partial^3 \Delta \tilde{G}_m}{\partial x^3} = 0 = \frac{1}{(1 - x_c)^2} - \frac{1}{x_c^2}. \tag{3.71}$$

We then obtain $B_c = \frac{1}{2}$ and $x_c = \frac{1}{2}$ as the critical values that separate the single phase from the binary phase regime in Fig. 3.3.

This very simple model provides a semiquantitative framework for understanding the origin of phase separation phenomena in real binary solutions. There have been many attempts to generalize the presentation by eliminating the strictly symmetric approach. A discussion of these features leads us too far afield.[3]

Thermodynamic representation

We now investigate how thermodynamics is used to characterize the approach to critical phenomena. We begin again with the so-called Bragg–Williams approximation

$$\frac{\Delta \tilde{G}_m}{RT} \equiv \frac{\tilde{G}_m - \tilde{G}^*}{RT} = (1 - x)\ln(1 - x) + x\ln x + \frac{w}{RT}x(1 - x). \tag{3.72}$$

3. The interested reader may consult J.M. Honig, *Thermodynamics* (Academic Press, Amsterdam, 2014), Secs. 3.8 and 3.9.

At temperatures $RT/w < \frac{1}{2}$, and in the range where the mixture splits into two phases of composition x' and x'', we need to determine the relative amounts of material in each phase. The mole numbers for component 2 at compositions $'$ and $''$ are given by

$$n_2 = n_2' + n_2'' \quad \text{or} \quad \frac{n_2}{n} = \left(\frac{n_2'}{n'}\right)\left(\frac{n'}{n}\right) + \left(\frac{n_2''}{n''}\right)\left(\frac{n''}{n}\right). \tag{3.73}$$

Let $f \equiv \frac{n''}{n}$ represent the fraction of material associated with the $''$ phase. On converting to mole fractions we then introduce the conservation law for component 2 as ($x_2 \equiv x$)

$$x = x'(1-f) + x''f, \tag{3.74}$$

which may be solved for

$$f = \frac{x - x'}{x'' - x'}, \tag{3.75}$$

in which x is the mole fraction of component 2 in the homogeneous phase. Equation (3.75) is known as the *lever rule*. Here x' or x'' are no longer variables under the control of the experimenter; rather, they are uniquely determined by the solution temperature.

We determine the Gibbs free energy of the binary phase by writing

$$G = G' + G'' \quad \text{or} \quad \tilde{G} = \frac{n'}{n}\tilde{G}' + \frac{n''}{n}\tilde{G}'' = (1-f)\tilde{G}' + f\tilde{G}''. \tag{3.76}$$

Then replace $\tilde{G}'$ and $\tilde{G}''$ by applying Eq. (3.72) to the two phases in equilibrium with each other:

$$\begin{aligned}\tilde{G} = {} & (1-f)\tilde{G}_1^* + f\tilde{G}_2^* + RT(1-f)\{(1-x')\ln(1-x') + x'\ln x'\} \\ & + RTf\{(1-x'')\ln(1-x'') + x''\ln x''\} \\ & + w(T)\{(1-f)x'(1-x') + fx''(1-x'')\}.\end{aligned} \tag{3.77}$$

Within the present approximations the minima in Fig. 3.3 are symmetrically displaced with respect to $x = 0$ and $x = 1$, whence $x' = 1 - x''$. Equation (3.77) then reduces to

$$\frac{\tilde{G}}{RT} = \frac{\tilde{G}^*}{RT} + \{(1-x'')\ln(1-x'') + x''\ln x''\} + \frac{w(T)}{RT}x''(1-x''). \tag{3.78}$$

Recall that $x''(T)$ is not arbitrarily adjustable but rather is specified by locating the minima of Eq. (3.72) when $\frac{RT}{w} < \frac{1}{2}$. Thus, Eq. (3.78) depends solely on T, in contrast to Eq. (3.72), where the composition variable x is under the direct control of the experimenter.

To show how $\tilde{G}$ varies with temperature, we need to determine the temperature dependence of x'' by minimizing Eq. (3.78). This requires a numerical solution of the transcendental equation (3.69). This operation is not of particular interest. It is of greater relevance to study how the enthalpy changes with temperature.

Proceeding as usual, the enthalpy is obtained via the thermodynamic relation

$$H = -RT^2 \left[\frac{\partial \left(\frac{G}{RT} \right)}{\partial T} \right]. \tag{3.79}$$

Using Eq. (3.78) in which, for $\frac{RT}{w} < \frac{1}{2}$, $x'' = x''(T)$ is an implicit function of temperature, the differentiation process yields

$$\begin{aligned} -\frac{\tilde{H}}{RT^2} = -\frac{\tilde{H}^*}{RT^2} + \left\{ \ln\left[\frac{x''}{1-x''}\right] + \frac{w(T)}{RT}(1-2x'') \right\} \frac{dx''}{dT} \\ + x''(1-x'')\left[\frac{1}{RT}\frac{dw(T)}{dT} - \frac{w(T)}{RT^2}\right]. \end{aligned} \tag{3.80}$$

On account of Eq. (3.69) the central term in braces drops out; Eq. (3.80) then collapses to read

$$\tilde{H} = (1-f)\tilde{H}_1^* + f\tilde{H}_2^* + \left[w(T) - T\frac{dw(T)}{dT}\right] x''(1-x''). \tag{3.81}$$

It is conventional at this point to introduce the *order parameter* s_p defined by

$$s_p \equiv 2x'' - 1 \quad \text{or} \quad x'' = \frac{1}{2}(1+s_p), \tag{3.82}$$

with which the equilibrium condition (3.69) assumes the simpler form

$$\ln\left[\frac{1+s_p}{1-s_p}\right] = \frac{w(T)}{RT} s_p. \tag{3.83}$$

Equations (3.78) and (3.81) now read

$$\begin{aligned} \Delta\tilde{G}_m = RT\left\{\frac{1}{2}(1+s_p)\ln(1+s_p) + \frac{1}{2}(1-s_p)\ln(1-s_p) - \ln 2\right\} \\ + \frac{w}{4RT}\left(1-s_p^2\right) \end{aligned} \tag{3.84}$$

and

$$\Delta\tilde{H}_m = \frac{1}{4}\left[w(T) - T\frac{dw(T)}{dT}\right](1-s_p^2). \tag{3.85}$$

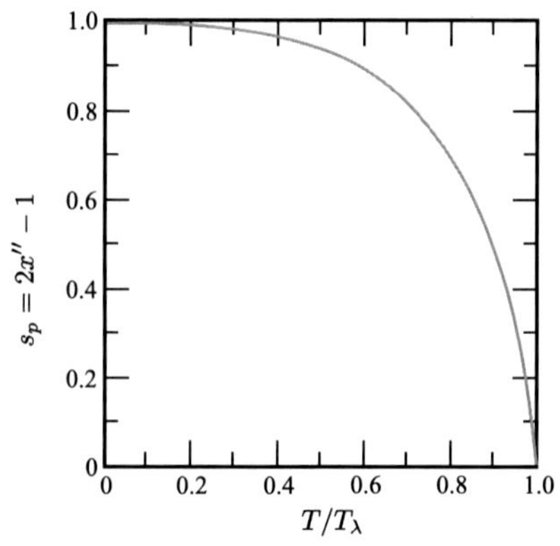

FIGURE 3.5 Change of the order parameter s_p with T/T_λ as determined by Eq. (3.87).

Equation (3.83) may be recast in the equivalent form

$$\tanh\left(\frac{ws_p}{2RT}\right) = s_p. \tag{3.86}$$

We now introduce a *characteristic temperature* by $T_\lambda \equiv \frac{w}{2R}$, usually coupled to the assumption that w is constant. Equation (3.86) then reads

$$\frac{T_\lambda}{T} = \frac{\tanh^{-1} s_p}{s_p}. \tag{3.87}$$

This finally paves the way to determining the variation of s_p with $\frac{T}{T_\lambda}$ via numerical solution of Eq. (3.87); the universal curve is depicted in Fig. 3.5. Of particular interest is the fact that close to criticality we may expand the inverse hyperbolic function as $\tanh^{-1} s_p \approx s_p + \frac{s_p^3}{3}$. Insertion into Eq. (3.87) yields

$$\frac{T}{T_\lambda} \approx 1 - \frac{s_p^2}{3} \quad \text{or} \quad s_p = \sqrt{3\left(1 - \frac{T}{T_\lambda}\right)} \quad (T \to T_\lambda^-). \tag{3.88}$$

We have encountered precisely this behavior pattern before: the order parameter approaches zero as $\sqrt{T_\lambda - T}$, which is characteristic of the mean field theory and identifies T_λ as the critical point. The order parameter exhibits a discontinuity in its first derivative but is continuous across T_λ; hence, we are dealing with a higher-order transition.

Using Fig. 3.5, we can determine the variation of the enthalpy change via Eq. (3.85). The result is depicted in Fig. 3.6 on neglecting the $\frac{dw(T)}{dT}$ term. The graph is continuous but exhibits a discontinuity in slope at the critical temperature.

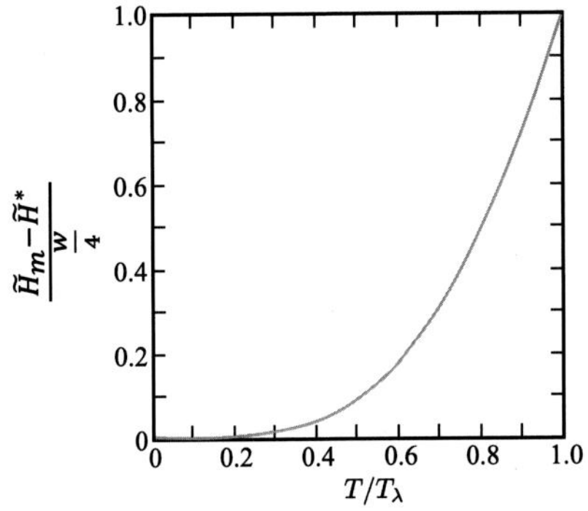

FIGURE 3.6 Change in the enthalpy of mixing as a function of T/T_λ as determined by Eq. (3.85) with $\frac{dw}{dT}=0$.

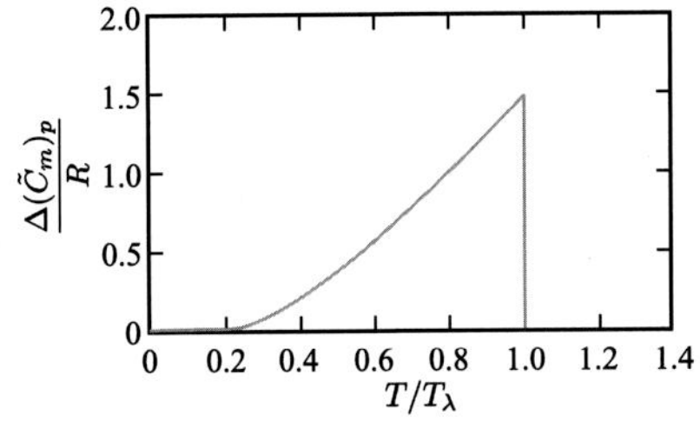

FIGURE 3.7 Plot of the scaled heat capacity vs. T/T_λ as determined from the slope of Fig. 3.6.

The heat capacity at constant pressure is found via the relation

$$\Delta C_m\bigg|_P = \left(\frac{\partial \Delta \tilde{H}_m}{\partial T}\right)_P . \tag{3.89}$$

The resulting curve with $w = 2R$ is shown in Fig. 3.7 as a plot of $\frac{\Delta C_m}{R}$ vs. $\frac{T}{T_\lambda}$. Consistently with mean field theory, we observe a discontinuity in the heat capacity at the critical temperature.

Discussion

Although the above discussion relates specifically to the thermodynamics of mixing, it can be applied to any case in which the elementary constituents of a system obey Eq. (3.62) and fall into one of two possible configurations: A or B, up or down, full or empty, plus or minus, and so on. This follows as an inescapable consequence from the modest beginnings of the Margules approach, which itself is consistent with the general principles of mean field theory that comes under the more general heading of *order–disorder theory*. We thus see again that mean field theory is not confined to lattice models with nearest-neighbor interactions.

APPENDIX 3.A USEFUL RELATIONS RELATED TO THE VAN DER WAALS EQUATION

We begin with the van der Waals equation of state $P = \frac{RT}{\tilde{V}-b} - \frac{a}{\tilde{V}^2}$ where $\tilde{V}$ is the molar volume of the fluid. Taking differentials, we obtain

$$dP = \left(\frac{R}{\tilde{V}-b}\right) dT + \left(2\frac{a}{\tilde{V}^3} - \frac{RT}{(\tilde{V}-b)^2}\right) d\tilde{V}, \tag{3.A.1}$$

from which we read off

$$\begin{aligned} \left(\frac{\partial P}{\partial T}\right)_V &= \frac{R}{\tilde{V}-b}; \\ \left(\frac{\partial \tilde{V}}{\partial T}\right)_P &= \frac{\frac{RT}{\tilde{V}-b}}{\frac{RT}{(\tilde{V}-b)^2} - \frac{2a}{V^3}}; \\ \left(\frac{\partial \tilde{V}}{\partial P}\right)_T &= \left[2\frac{a}{\tilde{V}^3} - \frac{RT}{(\tilde{V}-b)^2}\right]^{-1}. \end{aligned} \tag{3.A.2}$$

We also use the thermodynamic relation $\left(\frac{\partial C_V}{\partial V}\right)_T = T\left(\frac{\partial^2 P}{\partial T^2}\right)_V = 0$, which vanishes for the van der Waals equation of state. The fact that C_V is a constant independent of volume is characteristic of ideal gases as well, where $\tilde{C}_V = \frac{3R}{2}$ for monatomic species. Now set $\tilde{V} = \tilde{V}_c$, allowing the partial derivatives (3.A.2) to approach T_c; then insert (3.A.2) into Eq. (3.47). Finally, we need the standard relations that apply at the critical point: $a = \frac{9RT_c\tilde{V}_c}{8}$, $b = \frac{\tilde{V}_c}{3}$. On simplifying we recover Eq. (3.48).

Further, feeding this information into the defining relation for κ_T and after considerable simplification, we arrive at Eq. (3.52).

APPENDIX 3.B DEFINITION OF CORRELATION FUNCTIONS

The methodology for constructing correlation functions can also be used when no external field is present. In that case, we add a fictitious term in the form $-X_iY$ to the Hamiltonian, where X_i is a quantity whose average is to be determined, and Y is the conjugate variable. The energy values are denoted by E_i. Then execute the operation

$$\begin{aligned} \frac{1}{\beta}\left(\frac{\partial \ln Z}{\partial Y}\right)_{Y=0} &= \frac{1}{\beta Z}\left[\frac{\partial}{\partial Y}\sum_i e^{-\beta(E_i - X_iY)}\right]_{Y=0} \\ &= \frac{1}{\beta Z}\left[\sum_i X_i e^{-\beta(E_i - X_iY)}\right]_{Y=0} = \langle X_i \rangle, \end{aligned} \tag{3.B.1}$$

which provides an alternative method for determining expectation values. We leave to the reader to verify the following: Starting with the free energy function $F = -\frac{1}{\beta}\ln Z$ and executing a second differentiation with respect to Y establishes the *fluctuations*

$$\langle X_i^2 \rangle - \langle X_i \rangle^2 = \frac{1}{\beta}\frac{\partial \langle X_i \rangle}{\partial Y} \equiv \frac{\chi_i}{\beta}, \tag{3.B.2}$$

where χ is a generalized susceptibility. This introduces in general terms the linear response relationship introduced elsewhere.

APPENDIX 3.C GIBBS–DUHEM RULES

The Gibbs–Duhem equation is based on the relation between the Gibbs free energy and the chemical potentials of a mixture of n_i moles of components i, each with an associated chemical potential μ_{ci}, for which

$$G(T, P; n_1, n_2 \ldots, n_r) = \sum_{i=1}^{r} n_i \mu_{ci}(T, P). \tag{3.C.1}$$

Its differential

$$dG = \sum_{i=1}^{r} n_i d\mu_{ci} + \sum_{i=1}^{r} \mu_{ci} dn_i \tag{3.C.2}$$

may be compared with the standard form

$$\begin{aligned} dG &= \left(\frac{\partial G}{\partial T}\right)_{P,n_i} dT + \left(\frac{\partial G}{\partial P}\right)_{T,n_i} dP + \sum_{i=1}^{r}\left(\frac{\partial G}{\partial n_i}\right)_{T,P,n_j} dn_i \\ &= -SdT + VdP + \sum_{i=1}^{r} \mu_{ci} dn_i. \end{aligned} \tag{3.C.3}$$

Comparison of Eq. (3.C.3) with (3.C.2) shows that

$$\sum_{i=1}^{r} n_i d\mu_{ci} = -SdT + VdP, \tag{3.C.4}$$

which is known as the *Gibbs–Duhem* equation. At constant T and P Eq. (3.C.4) reduces to $\sum_{i=1}^{r} x_i d\mu_{ci} = 0$. The canonical relation for the chemical potential of component 1 is given by

$$\mu_{c1} = \left(\frac{\partial G}{\partial n_1}\right)_{T,P} = \mu_{c1}^{*} + RT\ln x_1. \tag{3.C.5}$$

In nonideal solutions x_i is replaced by a_i. Taking differentials at constant T and P and substituting into Eq. (3.C.4), we obtain the Gibbs–Duhem relation.

APPENDIX 3.D FORMULA FOR THE ENTROPY PART IN EQ. (3.61)

We note that the number of microconfigurations of a binary mixture is

$$\Omega = \frac{N!}{n_1!n_2!}, \tag{3.D.1}$$

where $N = n_1 + n_2$. Taking natural logarithm of Eq. (3.D.1) and using the first Stirling formula for large numbers (e.g., $\ln n_i! = n_i(\ln n_i - 1)$), we obtain the molar entropy as

$$S_m \equiv k_B N_{AV} \ln \Omega = R\,(n_1 \ln n_1 + n_2 \ln n_2)\,. \tag{3.D.2}$$

Taking $x_i \equiv n_i/N_{mol}$, N_{AV} as the Avogadro number, $R \equiv k_B N_{AV}$, and the entropy part in the molar Gibbs free energy as $(-T S_m)$, we arrive at the starting expression (3.61). Note that in that expression the internal energy part is regarded as a constant contained in the G^* reference free energy.

Chapter 4

The Landau Theory of Phase Transitions: General Concept and Its Microscopic Relation to Mean Field Theory

As a parting shot to mean field theory, we take up the Landau theory of phase transitions. This theory in its original formulation does not involve any specific physical models, as has been the case in Chapters 2 and 3, but is based entirely on the concepts of spontaneously broken symmetry and on the corresponding order parameter. It is thus capable of far-ranging generalizations, some of which we consider below. The general transparency of Landau's approach stands in distinct contrast to the mathematical intricacies that await us when we go beyond mean field theory. Also, Landau's approach will be seen to be extremely flexible and therefore worth further study, even if the final results agree only qualitatively, or at best semiquantitatively, with experiment.

4.1 CONCEPT OF THE ORDER PARAMETER

We first introduce the fundamental notion of an order parameter η as the entity that characterizes the degree of ordering when a phase transformation takes place on lowering, e.g., the temperature. As a familiar example, already mentioned earlier, consider the magnetization of a ferromagnetic domain, i.e., a region of aligned magnetic moments. At low temperatures $T \ll T_c$, all spins are perfectly aligned, which corresponds to taking $\eta = 1$, whereas for $T \geq T_c$, the thermal noise leads to a chaotic misalignment of the spins: random spin orientations prevail. This situation corresponds to taking $\eta = 0$. In the intermediate range $0 < T < T_c$, the thermal agitation is not completely destructive and allows for a partial average alignment of the spins. The magnetization (polarization of spins) thus changes gradually from totality to zero, which makes it natural to associate η with the magnetization M, as we did in Chapter 3. The experimental temperature dependence is sketched in, e.g., Fig. 2.4. Note in particular the vertical descent as $\eta \to 0$.

A Primer to the Theory of Critical Phenomena. **http://dx.doi.org/10.1016/B978-0-12-804685-2.00004-8**

Within the Landau approach, we are particularly interested in the study of a system close to criticality, as $T \to T_c^-$ and $\eta \to 0$. We then assume that, close to T_c, all thermodynamic properties may be represented in an ascending power series in η, which can be truncated. Thus, we write

$$\mathcal{L} = a_0 + a_1\eta + a_2\eta^2 + a_3\eta^3 + a_4\eta^4 + \cdots , \tag{4.1}$$

where $\mathcal{L}$ is the so-called *Landau free energy.*

This formal decomposition requires a careful additional comment. First, if we take $\eta = M$, then in an applied magnetic field, we have to add the term $-MH_a \equiv -\eta H_a$, but then a complication arises because $\mathcal{L}$ depends simultaneously on magnetization and on the applied field, the two thermodynamically conjugate variables. Therefore, strictly speaking, $\mathcal{L}$ is not a thermodynamic function of state. To address the situation, Landau introduced the concept of a generalized free (or Gibbs) energy, now called *the Landau free (or Gibbs) energy functional* $\mathcal{F}$, which contains extra terms. These depend on the additional variable (order parameter) η so that $\mathcal{L} \equiv \mathcal{L}(H_a, T; \eta)$. The true (physical) free energy is obtained by minimizing $\mathcal{L}$ with respect to η, i.e., using the condition[1]

$$\left.\frac{\partial \mathcal{L}}{\partial \eta}\right|_{\bar{\eta}} = 0, \quad \left.\frac{\partial^2 \mathcal{L}}{\partial \eta^2}\right|_{\bar{\eta}} > 0, \tag{4.2}$$

and substituting the resulting $\eta \equiv \bar{\eta}(T, H_a)$ into $\mathcal{L}$, so that

$$F(H_a, T) \equiv \mathcal{L}\,(H_a, T; \bar{\eta}(H_a, T))\,. \tag{4.3}$$

In that situation, $\bar{\eta}(H_a, T)$ represents the thermodynamically stable solution. This procedure represents the essence of the Landau approach that also requires a proper choice of the order parameter appropriate for describing a given physical situation.

A few additional and important remarks should be made at this stage.

Important note

First, the constant term a_0 is identified with $\eta \equiv 0$, the disordered thermodynamic state. Second, the expansion (4.1) contains only even-order terms, as it is assumed that the solution is symmetric with respect to sign of η. In other words, if we assume that η represents, e.g., the magnetization M, then for $H_a = 0$, the solutions $+\eta$ and $-\eta$ are equivalent. This expresses the general idea that, in the disordered state, both magnetization orientations $\pm\eta$ should be equivalent, i.e.,

---

1. Usually, the absolute value of the order parameter lies in the interval [0, 1]. Then, for example, we find that $\eta = M/S$. In such a situation, we have to check separately whether the minimum of $\mathcal{L}$ is located at the end values for η, since we determine the minimum of $\mathcal{L}$ in a finite interval.

the time reversal symmetry is not broken. In the ordered phase, only one solution is taken (say, $+\eta$), and this represents the idea of a spontaneous breakdown of the system symmetry as $T \to T_c^+$. Third, this is the reason why we consider the onset of ordering at T_c an *emergent phenomenon*. By that we mean that we have to assume first that $\eta \neq 0$ and then check whether this solution is thermodynamically stable, rather than the $\eta \equiv 0$ solution, which is always present if the external field is absent. All those features will be elaborated on in detail in the following section. If Landau's methodology is to be useful in describing the phenomenological aspects of phase transitions, then at least some of the expansion coefficients must be allowed to depend parametrically on the temperature and, if needed, on other variables, such as pressure or applied magnetic field. Such a dependence is introduced at a phenomenological level in an ad hoc fashion. A proper theoretical approach is based at the microscopic level appropriate to the prevailing physical situation.

Symmetry requirements

We now engage in a systematic discussion of Eq. (4.1) under a variety of choices for the a_i coefficients. The constant a_0, with dimensions of energy density, may be chosen to represent the background free energy density and may thus be included in $\mathcal{L}$, which then represents events beyond those of the background.

We enforce equilibrium conditions by minimizing Eq. (4.1):

$$\frac{\partial \mathcal{L}}{\partial \eta} = 0 = a_1 + 2a_2\eta_0 + 3a_3\eta_0^2 + 4a_4\eta_0^3 + \cdots . \tag{4.4}$$

Expansion in even powers of the order parameter; Higher-order transitions

We now examine a variety of special cases, the simplest of which invokes isotropy. Then $\mathcal{L}$ is an even function of η: there can be no external features that provide a direction to the approach to the critical point. Accordingly, in Eq. (4.1), we eliminate the terms involving odd powers of the order parameter, setting $a_1 = a_3 = 0$.

As already remarked, the coefficients a_2 and a_4 are to reflect the actions of the relevant control variables, which normally involve the temperature T. We therefore set $a_2 = a_2(T) \equiv \frac{1}{2}a \cdot (T - T_c) \equiv \frac{1}{2}at_c$, which provides a first-order linear temperature departure of the system from criticality. For convenience, we further set $a_4 \equiv \frac{1}{4}b$; here we neglect the temperature variation since it produces second-order effects that we ignore. Accordingly, the restricted Landau functional reads:

$$\mathcal{L} = \frac{1}{2}at_c\eta^2 + \frac{1}{4}b\eta^4. \tag{4.5}$$

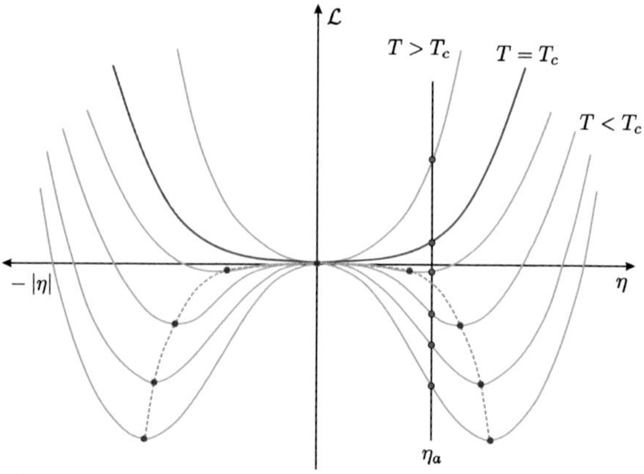

FIGURE 4.1 Schematic representation of Eq. (4.5). See the text for a description of the heavy dots.

Representative isotherm plots of $\mathcal{L}$ as a function of η for $t_c < 0$, $t_c = 0$, and $t_c > 0$ are shown in Fig. 4.1; these curves mimic those of Fig. 2.2. Note again the changeover from a single minimum in $\mathcal{L}$ (at $\eta = 0$ when $T \geq T_c$) to a double minimum (at $\eta = \pm\sqrt{\frac{-at_c}{b}}$ when $T < T_c$; see below). The flatness of the $T = T_c$ curve allows for the possibility of encountering large fluctuations.

If we introduce the thermodynamic representation in terms of changes in enthalpy ΔH_c and entropy ΔS_c (relative to the background) via $\mathcal{L} = \Delta H_c - T\Delta S_c$, then

$$\Delta H_c = -\frac{1}{2}aT_c\eta^2 + \frac{1}{4}b\eta^4; \quad \Delta S_c = -\frac{1}{2}a\eta^2. \tag{4.6}$$

At this point, we apply the equilibrium constraint $\frac{\partial \mathcal{L}}{\partial \eta} = 0$ that leads to

$$a(T - T_c)\eta_0 + b\eta_0^3 = 0, \tag{4.7}$$

thereby introducing constraints on the order parameter. This relation has three solutions

$$\eta_0 = 0 \; (T \geq T_c) \quad \text{or} \quad \eta_0 = \pm\sqrt{\frac{a(T_c - T)}{b}} \equiv \pm\left(\frac{-at_c}{b}\right)^{\frac{1}{2}} \; (T < T_c), \tag{4.8}$$

which link the order parameter to temperature; we discard the negative solution as unphysical. Above the critical temperature, η_0 becomes complex; this forces us to adopt the solution $\eta_0 = 0$. To keep η_0 real, we require that $\frac{a}{b} > 0$; also, in setting up Eq. (4.5), we assumed that $a > 0$, thus rendering $b > 0$. Slightly below criticality, η_0 varies as $\sqrt{T_c - T}$. This is precisely the T dependence of the order parameter we encountered in mean field theory. Again, the experimental exponent is closer to the value $\beta = \frac{1}{3}$. Furthermore, this approach is again in agreement with the experimental fact that the critical temperature is approached with infinite slope.

Contact with thermodynamics is achieved by inserting Eq. (4.8) into Eq. (4.5); then the stationary solution becomes the Gibbs free energy[2] with the quadratic form

$$\Delta G = 0 \ (t_c \geq 0) \ \text{and} \ \Delta G = -\frac{1}{4}\frac{a^2 t_c^2}{b} \quad (t_c < 0). \tag{4.9}$$

The free energy in excess of the background thus vanishes at and above the critical point and is a parabolic function of the temperature below T_c.

The entropy is determined as

$$\Delta S = 0 \ (T \geq T_c) \ \text{or} \ \Delta S = -\left(\frac{\partial \Delta G}{\partial T}\right) = \frac{a^2}{2b} t_c \quad (T < T_c). \tag{4.10}$$

The entropy ΔS is negative since increasing spin ordering at lower temperatures reduces the entropy of the system. Note that ΔS changes smoothly in crossing the critical temperature, whereas the first derivative is discontinuous. This is indicative of a higher-order transition.

The excess heat capacity at constant pressure can be worked out from the thermodynamic relation $\frac{\Delta C_P}{T} = \left(\frac{\partial \Delta S}{\partial T}\right)_P$, which leads to

$$C_P = 0 \ (T \to T_c^+) \ \text{and} \ C_P = \frac{a^2}{2b} T \quad (T \to T_c^-). \tag{4.11}$$

In contrast to the function ΔG, C_P exhibits a discontinuity at the transition point. This discontinuity was also encountered previously in mean field theory (cf. Fig. 3.7). However, experimentally a singularity in ΔC_P is observed at the critical temperature.

The above mathematical operations are perhaps best illustrated by referring to Fig. 4.1. In principle, we can ask how, for a fixed value η_a of the order parameter, the Landau functional changes with temperature; this is indicated by the heavy dots on the vertical line in Fig. 4.1. However, under equilibrium conditions the order parameter itself changes with temperature according to Eq. (4.8). This leads to the progression of heavy dots along the minima as the only stable configuration, corresponding to Eq. (4.9). Note an interesting feature of relevance to our future discussion, by reverting back to the unrestricted order parameter. The Landau functional should always rise with increasing absolute

2. In dealing with condensed phases, we ignore the small difference between the Gibbs and Helmholtz free energies, so that constant pressure rather than constant volume conditions may be assumed to prevail.

values of the order parameter since the entropy diminishes and the enthalpy rises with increasing ordering. This is the case outside the dome surrounding the locus of the equilibrium minima; however, with increasing $|\eta|$ the opposite is encountered within the dome. Thus, all configurations within the dome are unstable, and the corresponding η values are not encountered under stable conditions. This observation has its counterpart in thermodynamics under the heading of the "common tangent construction." Whereas under equilibrium conditions the free energy minima are equal at fixed temperatures, the state of the system changes discontinuously from $\Delta G(T, -|\eta_0|)$ to $\Delta G(T, \eta_0)$, thus eliminating all η values between the minima.

Effects in applied magnetic field

We now introduce a greater degree of flexibility by including the effect of an externally applied field, such as the magnetic field B. The linear response to it is the "magnetization" (density) $\langle S^z \rangle \equiv M \equiv \eta$, which continues to serve as the relevant order parameter. We deal here with the particular case where the magnetization is aligned with the imposed field, leading to the additional contribution $-\mu BM$ to the energy of the system. This introduces a spatial anisotropy, whereby the Landau functional reads

$$\mathcal{L} = -\mu B\eta + \frac{1}{2}at_c\eta^2 + \frac{1}{4}b\eta^4, \tag{4.12}$$

where we have already included the temperature dependence of a_2. Obviously, the subsequent analysis is not limited to magnetization effects; any system that includes a linear order parameter dependence is handled in the same manner.

In surveying the ramifications of Eq. (4.12) we follow the exposition of Goldenfeld.[3] The dependence of $\mathcal{L}$ on η is discussed in relation to Fig. 4.2. The three columns left to right depict the variation of $\mathcal{L}$ with η in its dependence on decreasing temperature when the magnetic field is held at $B < 0$, $B = 0$, $B > 0$ respectively; the three rows top to bottom depict the $\mathcal{L}(\eta)$ dependence on B for the cases $T > T_c$, $T = T_c$ and $T < T_c$ respectively. The heavy dots locate the equilibrium values of η.

As expected, we now encounter a richer set of variations of $\mathcal{L}(\eta)$. When $T \geq T_c$, and for $B = 0$ the free energy at equilibrium in excess of its background value occurs at the single minimum with $\eta = 0$. As we move left to right and down along the first two rows, the shift from negative to positive B is accompanied by a continuous shift in η past zero at $B = 0$, toward more positive values. We have encountered an expected dependence of the order parameter on the magnetic field.

3. N. Goldenfeld, *Lectures on Phase Transitions and the Renormalization Group*, Perseus Books, Reading, MA, 1992, pp. 141 ff.

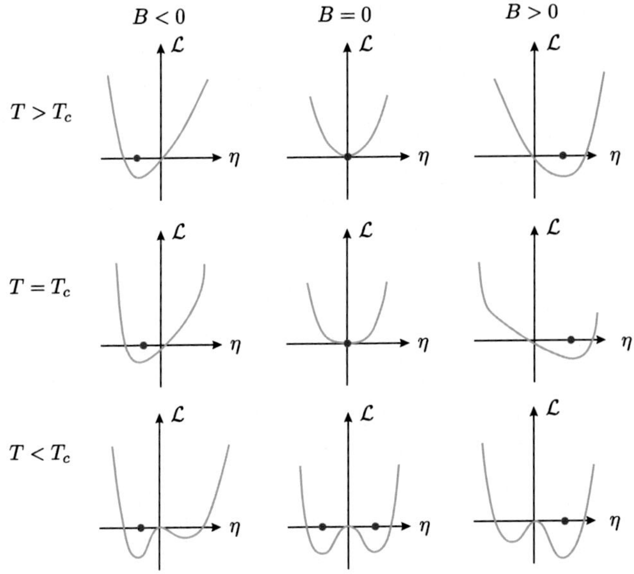

FIGURE 4.2 Diagrams sketching the change of the Landau functional with magnetic field and temperature as discussed in the text. Adapted from N. Goldenfeld, footnote 3.

In the bottom row, for $T < T_c$ and $B < 0$, $\mathcal{L}(\eta)$ displays a double minimum which is asymmetric in the presence of the field; the equilibrium configuration obviously coincides with the lower of the two minima. When $B = 0$ the two equivalent minima correspond to the same $\mathcal{L}$ value; thermodynamics dictates that the intervening η range is inaccessible. For $B > 0$, the system exhibits properties corresponding to the lower minimum on the right.

The central column shows what happens as T is decreased when the applied field is absent. This summarizes the preceding discussion and the results of Chapter 2. The single value of $\eta = 0$ at the $\mathcal{L}$ minimum is replaced by two minima at nonzero η values when the temperature is lowered past the critical value: now, at any given temperature in the range $T < T_c$, intermediate values of η are not encountered.

Magnetic field effects

We now consider the analytical representation when $B \neq 0$. Imposing the requirement $\frac{\partial \mathcal{L}}{\partial \eta} = 0$ on Eq. (4.12) leads to

$$a\, t_c\, \eta_0 + b\eta_0^3 = a\, t_c\, M + bM^3 = \mu B. \tag{4.13}$$

A similar approach can be applied to van der Waals fluids. We can ask again how it is that ferromagnets and van der Waals fluids can be treated by the same theoretical technique. This hinges on the fact that the relation $\pi = 1 + 4t_c + 6t_c\eta + \frac{3}{2}\eta^3$, (Eq. (3.42)), may be derived from the following free energy expression: $\mathcal{L} = \frac{1}{\rho_c^2}\left[-(\pi - 1 - 4t_c)\eta + 3t_c\eta^2 + \frac{3\eta^4}{8}\right]$. Comparison with Eq. (4.13) shows that $(\pi - 1 - 4t_c)$ plays the role of μB.

In the absence of a magnetic field, Eq. (4.13) reduces to

$$\eta_0 = 0 \text{ or } \eta_0 = \pm\sqrt{\frac{at_c}{b}}, \tag{4.14}$$

as before. Now, however, we will have to retain both sign options: in our rigid framework, as the field direction is reversed, the magnetization—hence the order parameter—also changes sign, so as to maintain its alignment with the field. Hence μB is always positive.

With the field present above the critical point, $t_c > 0$, $\eta_0 = 0$ is still an applicable solution. At the critical point, $t_c = 0$, so that, from Eq. (4.13),

$$\mu B = bM^3 \text{ or } M \propto B^{\frac{1}{3}}. \tag{4.15}$$

In agreement with mean field theory, at the critical temperature the magnetization obeys a power-law dependence on the applied field with $\delta = 3$.

We study magnetization effects via the isothermal magnetic susceptibility $\chi_T = \frac{\partial \mathcal{M}}{\partial B}$, where $\mathcal{M} \equiv N\mu^2 M$ is the total magnetization. This is easily found by differentiating Eq. (4.13) with respect to B:

$$(at_c + 3b\eta_0^2)\frac{\partial M}{\partial B} = \mu. \tag{4.16}$$

In very weak magnetic fields the isothermal magnetic susceptibility is then specified by

$$\chi_T = \left(\frac{\partial N\mu^2 M}{\partial B}\right)_T = \frac{N\mu^2}{at_c + 3b\eta_0^2}, \tag{4.17}$$

where η_0 is the solution of Eq. (4.13). Then:

(i) for $t_c > 0$,

$$\eta_0 = 0 \text{ and } \chi_T = \frac{N\mu^2}{at_c}; \tag{4.18}$$

(ii) for $t_c < 0$,

$$\eta_0^2 = -\frac{at_c}{b} \text{ and } \chi_T = -\frac{N\mu^2}{2at_c}. \tag{4.19}$$

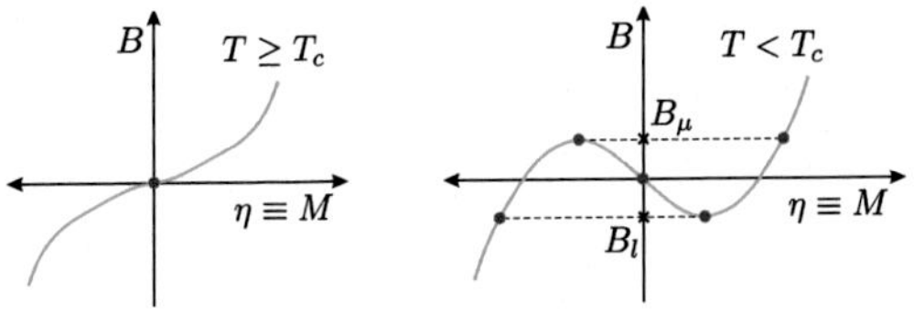

FIGURE 4.3 Diagrams illustrating the origin of magnetic hysteresis effects in Landau theory. See the text for a description.

Note that we have introduced an approximation by the use of Eq. (4.14). Equations (4.18)–(4.19) represent the Curie–Weiss law. Further, in units of $N\mu^2$, $\chi_T^{-1} = at_c$ above T_c and $\chi_T^{-1} = -2at_c$ below T_c. The two coincide at $t_c = 0$, whereas the first derivative is discontinuous, both indicative of a higher-order transition as χ_T^{-1} passes the critical temperature. This is confirmed by noting that the entropy of a magnetized system is proportional to $\left(-\frac{\partial \chi_T}{\partial t_c}\right)$.

The two versions (4.18) and (4.19) differ in sign and by a factor of two. Comparison with the mean field results indicates that $\gamma = 1$. This is a very elegant method of duplicating the rather cumbersome approach of Chapter 3.

Hysteresis effects via Landau theory

The above approach also provides a qualitative representation of hysteresis effects. Fig. 4.3 shows sketches of Eq. (4.13) for B as a function of η. Above the critical temperature, with $t_c > 0$, the plot shows a simple sigmoidal curve. Below T_c, with $t_c < 0$, a more complex curve provides for the possibility of encountering the same magnetic field B_u or B_l for two values of the order parameter. By analogy to our earlier discussion, the intermediate magnetizations are unstable, giving rise to hysteresis effects that can be visualized by rotating the diagram counterclockwise by 90° and then by 180° about the η axis.

Pressure–volume variations near the critical point

We briefly, and in a rather cavalier fashion, adapt the above to the case of fluids. Just as the magnetization is the order parameter corresponding to the magnetic field, the (excess relative) volume $\mathcal{V}$ of a liquid serves as the order parameter corresponding to the (scaled) pressure $p \equiv \frac{P-P_c}{P_c}$. This allows us to adopt the replacements $B \to p$ and $(\eta = M) \to (\eta = \mathcal{V})$. On applying the indicated substitutions we arrive at

$$p = b\mathcal{V}^3, \tag{4.20}$$

which is equivalent to Eq. (4.15), and

$$\mathcal{V} = \sqrt{-\frac{at_c}{b}}, \tag{4.21}$$

which is equivalent to Eq. $\left(\frac{\partial \mathcal{V}}{\partial p}\right)_T = \kappa_T \mathcal{V} = (at_c)^{-1}$ for $t_c > 0$ or $(-2at_c)^{-1}$ for $t_c < 0$, equivalent to Eqs. (4.18)–(4.19).

The reader may check that the heat capacity results agree with those of Eq. (4.11).

We see then that all results obtained in the mean field approximation of critical phenomena have been reproduced above with far less effort.

Adaptation to the first-order phase transitions

We now include a cubic term in the representation of the Landau free energy, whereby we write

$$\mathcal{L} = \underset{\text{(I)}}{\frac{1}{2}at_c\eta^2} + \underset{\text{(II)}}{\frac{1}{4}b\eta^4} - \underset{\text{(III)}}{\frac{1}{3}c\eta^3} \quad (a > 0, b > 0, c > 0). \tag{4.22}$$

The Roman numerals serve for identification purposes.

A series of representative sketches of $\mathcal{L}$ vs. η are shown in Fig. 4.5 for a set of decreasing temperatures going top to bottom. It is seen that, at high temperatures, we encounter a nearly parabolic curve with a specific minimum value $\mathcal{L}_s$, characteristic of state A. On lowering the temperature the negative term (III) becomes noticeable, resulting in a secondary minimum $\mathcal{L}_1$, which drops further until it coincides with $\mathcal{L}_s$ at T_m, where the two minima match. As shown below, the order parameter jumps discontinuously from $\eta = 0$ to a positive value, indicative of a first-order transition. With decreasing temperature in the range $T < T_m$, we deal with the system in state B. At T_c, term (I) changes sign, and at lower temperatures the upper minimum becomes irrelevant.

For a quantitative analysis, we impose the equilibrium requirement

$$b\eta_0^3 - c\eta_0^2 + a(T - T_c)\eta_0 = 0, \tag{4.23}$$

a cubic equation with the solutions

$$\begin{aligned} \eta_0 = 0 \quad \text{and} \quad \eta_0(T) &= \frac{c}{2b} \pm \sqrt{\left(\frac{c}{2b}\right)^2 - \frac{a(T - T_c)}{b}} \\ &\equiv \frac{c}{2b} \pm \sqrt{\left(\frac{c}{2b}\right)^2 - \frac{T - T_c}{T_1}}, \end{aligned} \tag{4.24}$$

where we have set $T_1 \equiv \frac{b}{a}$, which continues to keep the last term under the square root dimensionless.

The subsequent analysis is a bit tricky. To keep η_0 real, the sum under the square root must remain positive. When $T > T_c$, the largest T value satisfying

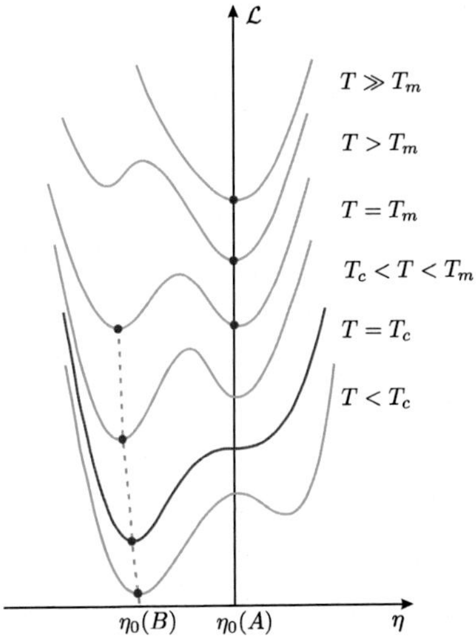

FIGURE 4.4 Schematic representation of Eq. (4.22). Note the first-order transition at temperature T_m from state A to state B.

this requirement is

$$T_m = T_c + T_1 \left(\frac{c}{2b}\right)^2, \tag{4.25}$$

which is greater than T_c. Note that

$$\eta_0(T_m) = \frac{c}{2b}; \quad \eta_0(T_c) = 0 \text{ or } \eta_0(T_c) = \frac{c}{b}. \tag{4.26}$$

As a second condition, we impose the requirement that η_0 remains nonnegative. In the range $T < T_c$, the square root in Eq. (4.24) always outweighs the first; this forces us to adopt the positive option.

The principal point of interest is that, in contrast to the case where the (III) term is missing, positive η_0 values extend into a temperature range above T_c; T_m is the first-order transition temperature; and T_1 is not an independent parameter, as Eq. (4.25) shows. Note that whereas η_0 diminishes with rising T when $T > T_c$, η_0 increases with T in the $T < T_c$ regime, which goes counter the perception that rising temperatures invariably increase disorder. Nevertheless, as Fig. 4.4 indicates, unless b is enormously greater than a or c, the negative terms (I) and (III) increasingly outweigh the positive (II) contribution, leading to a lowering of the lower minimum and rendering the equilibrium value of $\mathcal{L}$ more negative. This still leaves open the question how the mathematical analysis should be extended in the temperature range below T_c.

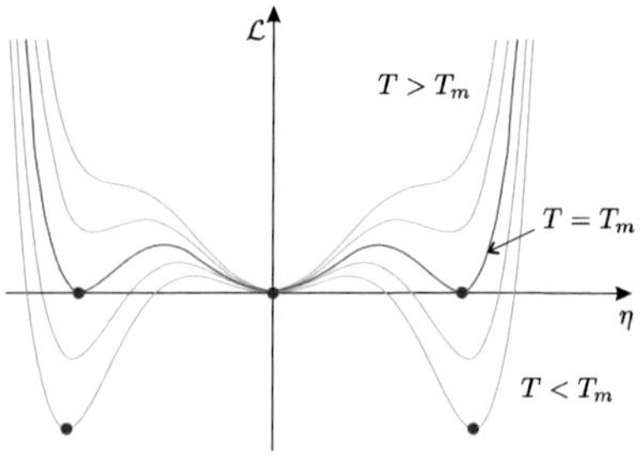

FIGURE 4.5 Schematic representation of Eq. (4.27). Note the first-order transition at temperature T_m.

A further quantitative assessment runs into complications because the truncated expansion of the order parameter may no longer be adequate for representing η.

A more complex Landau functional

This relates to the use of the functional

$$\mathcal{L}(\eta; T) = \frac{1}{2}a \cdot (T - T_c)\eta^2 - \frac{1}{4}b\eta^4 + \frac{1}{6}c\eta^6 \quad (a, b, c > 0), \tag{4.27}$$

where the coefficient of the quartic term is negative. The resulting $\mathcal{L}(\eta; T)$ curves are shown in Fig. 4.5. We are back to a symmetric set of curves. With decreasing temperature, $\mathcal{L}(\eta; T)$ exhibits two identical minima centered on two distinct η values, indicative of a transition.

Equilibrium is enforced by minimization, which leads to the expression

$$a \cdot (T - T_c)\eta_0 - b\eta_0^3 + c\eta_0^5 = 0. \tag{4.28}$$

The solutions are

$$\eta_0 = 0 \text{ or } \eta_0^2 = \frac{1}{2c}\left[b \pm \sqrt{b^2 - 4ac \cdot (T - T_c)}\right], \tag{4.29}$$

which, when substituted into Eq. (4.27), show how for a given set of parameters a, b, and c, the free energy varies with T.

We now examine the case $T < T_c$, for which we rewrite the above as

$$\eta_0 = 0 \text{ or } \eta_0^2 = \frac{1}{2c}\left[b \pm \sqrt{b^2 + 4ac \cdot |T - T_c|}\right]. \tag{4.30}$$

To keep η_0^2 positive, we must select the positive sign.

The least real value for η_0 in Eq. (4.30) is achieved at the temperature $T_1 = T_c + \frac{b^2}{4ac}$, where the square root in (4.29) vanishes. At higher temperatures,

η_0^2 becomes complex, forcing us to use the root $\eta_0 = 0$. Thus, the transition occurs at temperature T_1, where η_0 changes discontinuously from 0 to $\sqrt{\frac{b}{2c}}$; we have encountered a first-order transition. It is easy to check that at T_c, $\eta_0 = \sqrt{\frac{b}{c}}$. The corresponding entropy change at T_1 is $\Delta S_1 = -\frac{a\eta_0^2}{2} = -\frac{ab}{2c}$.

Summary

The Landau approach certainly represents a very compact approach to the general theory of phase transitions, of which the above is a sampling. On your own, you can explore further ramifications by adopting different power series to express $\mathcal{L}$ in terms of η. The flexibility of the Landau theory is certainly impressive. However, it is subject to limitations: the expansion of $\mathcal{L}$ in a limited power series in η, and the failure to deal with fluctuation effects.

Exercise. Add η^6 to the Landau functional. What will happen?

4.2 SPIN MAGNETISM: MICROSCOPIC DERIVATION OF THE LANDAU THEORY WITH SPATIAL FLUCTUATIONS OF THE ORDER PARAMETER*

In describing spin magnetism, we should first determine the structure of the spins (or equivalently, of the moments) in the ground state, which represents a broken-symmetry state with respect to time reversal since the magnetic state has a nonzero value of $\langle \boldsymbol{S}_i \rangle = \langle 0|\boldsymbol{S}_i|0\rangle$. We have already considered simple ferromagnetic and two-sublattice antiferromagnetic states. Now, we introduce a more involved description of magnetic ordering, starting from the concept of an inhomogeneous molecular (effective) field.[4] Such a generalization of the effective field concept leads to additional terms in the Landau functional, among them an all-important order-parameter gradient term, which is needed in later discussions of fluctuations.

4.2.1 The Concept of Exchange Fields and Ferromagnetism

We start again from the spin Hamiltonian in an applied field[4] $\mathbf{H}_a$, which we rewrite in the following form:

4. The external applied field $\boldsymbol{H}_a \equiv \frac{1}{\mu_0}\nabla \times \boldsymbol{A}$ acts also on the *orbital degrees of freedom* via the momentum operator $\boldsymbol{p} \to (\boldsymbol{p} - e\boldsymbol{A})$. All $\boldsymbol{p} - e\boldsymbol{A}$ quantities are in SI units. Here we neglect the orbital part since we deal only with the case of localized spins.

$$\begin{aligned}\hat{\mathcal{H}} &= -\frac{1}{2}\sum_{i\neq j} J_{ij}\,\hat{\boldsymbol{S}}_i\cdot\hat{\boldsymbol{S}}_j - g\mu_B\,\mathbf{H}_a\cdot\sum_i\hat{\boldsymbol{S}}_i \\ &= -\sum_i g\mu_B\,\hat{\boldsymbol{S}}_i\left[\frac{1}{g\mu_B}\sum_{j(i)} J_{ij}\,\hat{\boldsymbol{S}}_j + \mathbf{H}_a\right] \\ &\equiv -g\mu_B\sum_i \hat{\boldsymbol{S}}_i\cdot\hat{\mathbf{H}}_i \equiv -\sum_i \hat{\mathbf{M}}_i\cdot\hat{\mathbf{H}}_i,\end{aligned} \tag{4.31}$$

where we introduced the local effective field operator[5]

$$\hat{\mathbf{H}}_i \equiv \frac{1}{g\mu_B}\sum_{j(i)} J_{ij}\,\hat{\boldsymbol{S}}_j + \mathbf{H}_a \tag{4.32}$$

and the magnetic moment operator

$$\hat{\mathbf{M}}_i \equiv g\mu_B\,\hat{\boldsymbol{S}}_i. \tag{4.33}$$

The sum on the r.h.s. of (4.32) involves a fixed number of spins (in the simplest situation involving z nearest neighbors). This summation requires some type of averaging; hence, we introduce the *mean-field approximation* by setting

$$\sum_{j(i)} J_{ij}\,\hat{\boldsymbol{S}}_j \approx \left\langle \sum_{j(i)} J_{ij}\,\hat{\boldsymbol{S}}_j \right\rangle = \sum_{j(i)} J_{ij}\,\langle \boldsymbol{S}_j\rangle, \tag{4.34}$$

where the averaging involves the equilibrium state. This leads to an effective field of the form

$$\hat{\mathbf{H}}_i = \mathbf{H}_i = \frac{1}{g\mu_B}\sum_{j(i)} J_{ij}\,\langle \boldsymbol{S}_j\rangle \,+\, \mathbf{H}_a. \tag{4.35}$$

When the external field is absent ($H_a = 0$), the above field is called the *molecular field*.

The introduction of the molecular field is equivalent to approximating the starting Hamiltonian in the following manner:

$$\hat{\mathcal{H}} \simeq -g\mu_B\sum_i \hat{\boldsymbol{S}}_i\cdot\mathbf{H}_i, \tag{4.36}$$

i.e., it assumes the single-site form.

The Hamiltonian (4.36) can be solved easily since it describes a lattice of spins, each in an effective field. Thus, we write the partition function for the

5. Note that each spin $\boldsymbol{S}_i$ appears twice in the exchange interaction; hence the factor $\frac{1}{2}$ is eliminated.

spin $\hat{\boldsymbol{S}}_i$ explicitly in the form

$$\begin{aligned} Z_i &= \sum_{S_i^z} \exp\left(-\beta g \mu_B \langle S_i^z \rangle H_i\right) \\ &= \exp\left(-\beta g \mu_B S H_i\right) + \exp\left[-\beta g \mu_B (S-1) H_i\right] + \cdots \\ &\quad + \exp\left(+\beta g \mu_B S H_i\right) \\ &= \exp\left(+\beta g \mu_B S H_i\right) \frac{1 - \exp\left[-\beta g \mu_B (2S+1) H_i\right]}{1 - \exp\left(-\beta g \mu_B H_i\right)}, \end{aligned} \tag{4.37}$$

where the direction of the mean field serves as the spin quantization axis. Here we have used the partial summation of a geometric progression in the form

$$\sum_{m=-S}^{S} e^{-xm} = e^{+xS} \frac{1 - e^{-x(2S+1)}}{1 - e^{-x}} \equiv \frac{\sinh\left[x\left(S + \frac{1}{2}\right)\right]}{\sinh\left(\frac{x}{2}\right)} \tag{4.38}$$

with $x \equiv \beta g \mu_B H_i$. Next, we calculate the free energy according to

$$F = -k_B T \ln Z = -k_B T \ln \prod_{i=1}^{N} Z_i = -k_B T \sum_{i=1}^{N} \ln Z_i. \tag{4.39}$$

In the simplest situation the effective field $H_i = H$ is the same on every lattice site, e.g., when the system orders ferromagnetically. In that situation, $Z_i \equiv Z$ and

$$F = -k_B T \ln Z = -N k_B T \ln \frac{\sinh\left[x\left(S + \frac{1}{2}\right)\right]}{\sinh\left(\frac{x}{2}\right)}. \tag{4.40}$$

Now we take the most important step after the introduction of the mean field, which is the use of the *self-consistency condition*, whereby the thermodynamic relation for the total magnetization is found to be

$$\mathcal{M}_i = -\frac{\partial F}{\partial H_a} = -\frac{\partial F}{\partial H_i} = -\beta g \mu_B \frac{\partial F}{\partial x}. \tag{4.41}$$

This yields the relation

$$\langle S_i^z \rangle \equiv S B_S \left[\beta S \sum_{j(i)} J_{ij} \langle S_j^z \rangle + \beta g \mu_B H_a S \right]. \tag{4.42}$$

The function

$$B_S(y) \equiv \frac{2S+1}{2S} \coth\left(\frac{2S+1}{2S} y\right) - \frac{1}{2S} \coth\left(\frac{y}{2S}\right), \tag{4.43}$$

where $y \equiv \beta S \sum_{j(i)} J_{ij} \langle S_j^z \rangle + \beta g \mu_B H_a S$ is called the *Brillouin function*. We see that Eq. (4.42) represents a system of self-consistent equations for the expectation values $\langle S_i^z \rangle$ of the spin moments.

4.2.2 The Appearance of the Gradient Term

Equation (4.41) becomes an analytic function in certain limits. Suppose that the ferromagnetic system is weakly inhomogeneous on the magnetic scale, i.e., $|\langle S_j^z \rangle - \langle S_i^z \rangle| \ll S$. Then we can represent the lattice by a quasi-continuous medium by setting

$$\begin{aligned}\sum_{j(i)} J_{ij} \langle S_j^z \rangle &\equiv \sum_{j(i)} J_{ij} \left(\langle S_j^z \rangle - \langle S_i^z \rangle \right) + \langle S_i^z \rangle \sum_{j(i)} J_{ij} \\ &\approx \sum_{j(i)} J_{ij} \left\{ \boldsymbol{R}_{ji} \cdot \nabla \langle S^z(\boldsymbol{r}) \rangle \Big|_{\boldsymbol{r}=\boldsymbol{R}_i} + \frac{1}{2} (\boldsymbol{R}_{ji} \cdot \nabla)^2 \langle S^z(\boldsymbol{r}) \rangle \Big|_{\boldsymbol{r}=\boldsymbol{R}_i} \right\} \\ &\quad + \langle S^z(\boldsymbol{r}) \rangle J_0, \end{aligned} \tag{4.44}$$

where $\boldsymbol{R}_{ji} = \boldsymbol{R}_j - \boldsymbol{R}_i$ is the distance from atomic site i to j, $J_0 \equiv \sum_{j(i)} J_{ij}$, and $\langle S^z(\boldsymbol{r}) \rangle$ represents average spin density.[6] If the system displays inversion symmetry the term involving ∇ drops out, and we are left with

$$g\mu_B (H - H_a) = \sum_{j(i)} J_{ij} \langle S_j^z \rangle \simeq J_0 \langle S^z(\boldsymbol{r}) \rangle + \frac{1}{2} \sum_{j(i)} J_{ij} \left(\mathbf{R}_{ji} \cdot \nabla \right)^2 \langle S^z(\boldsymbol{r}) \rangle \Big|_{\boldsymbol{r}=\mathbf{R}_i}. \tag{4.45}$$

This expression provides a continuum representation of the molecular field. For the case of nearest-neighbor interactions and for a simple cubic system, we obtain

$$\sum_{j(i)} J_{ij} \langle S_j^z \rangle \simeq J_0 \langle S^z(\boldsymbol{r}) \rangle + J_0 a_0^2 \nabla^2 \langle S^z(\boldsymbol{r}) \rangle \Big|_{\boldsymbol{r}=\mathbf{R}_i}, \tag{4.46}$$

where a_0 is the lattice parameter. We have thereby introduced a gradient term that assumes an important role in the further development of the theory of critical phenomena. We see that in general the effective field acting on given spin involves a correction that derives from its nonlocal nature, in the sense that the neighboring spins contribute to the field.

Weak inhomogeneities are expected to occur close to the phase transition from the ferromagnetic to the paramagnetic state, i.e., when $\langle S_i^z \rangle \to 0$. In these

6. To a good approximation, this quantity may be identified with the spin moment per unit cell (see below).

circumstances we may expand the Brillouin function using the relation

$$B_S(x) \simeq \frac{S+1}{3S}x. \tag{4.47}$$

In effect, the self-consistent equation for $\langle S^z(\boldsymbol{r})\rangle|_{\boldsymbol{r}=\mathbf{R}_i}$ takes the form

$$\begin{aligned}\langle S^z(\boldsymbol{r})\rangle &= \frac{1}{3}J_0\,\beta S(S+1)\left\langle S^z(\boldsymbol{r})\right\rangle + \frac{1}{3}\beta g\mu_B\,H_a S(S+1)\\ &\quad + \frac{1}{3}J_0\,a_0^2\,\beta S(S+1)\nabla^2\langle S^z\rangle\Big|_{\boldsymbol{r}=\boldsymbol{R}_i}.\end{aligned} \tag{4.48}$$

If the gradient term is neglected (i.e., when the spins respond uniformly to the applied magnetic field), we obtain the so-called *Curie–Weiss* law for the sample magnetization

$$M = g\mu_B\left\langle S^z(\boldsymbol{r})\right\rangle N = \frac{C_M}{T-\Theta}H_a \tag{4.49}$$

with the Curie constant given by

$$C_M \equiv \frac{1}{3k_B}(g\mu_B)^2 S(S+1)N, \tag{4.50}$$

and with the so-called paramagnetic *Curie temperature* in the form

$$\Theta \equiv \frac{1}{3k_B}J_0\,S(S+1). \tag{4.51}$$

We should note that for $H_a \to 0$, $\langle S^z\rangle \to 0$. Hence, this approximation does not account for spontaneous spin ordering in the absence of an external field. Equation (4.51) describes the polarization in the high temperature regime $T \to \infty$ ($T \gg \Theta$) because, as $T \to \Theta$, the magnetic susceptibility diverges: $\chi = \frac{M}{H_a} \to \infty$; then an infinitely small field can polarize the system. Therefore, to describe the appearance of a spontaneous magnetic moment, we require a better approximation than Eq. (4.47). This will be dealt with in the next section.

Note the following interesting feature of the present formulation: Eq. (4.48) may be written in the form

$$C\nabla^2 M(\boldsymbol{r}) - (T-\Theta)M(\boldsymbol{r}) + C_M\,H_a = 0, \tag{4.52}$$

where $C \equiv (1/3k_B)J_0\,a_0^2\,S(S+1)$ is the *exchange stiffness constant*. This equation is a linearized version of the Ginzburg–Landau equation for the space profile of the static magnetization. In this case, the only solution for $H_a = 0$ in the unbounded medium is $M(\boldsymbol{r}) = 0$.

4.2.3 Formal Equivalence of the Landau Expansion and the Mean-field Approach

The *mean-field approximation* can be written in terms of the Hartree–Fock approximation already introduced in the discussion for itinerant spins. Recall the latter approximation for the operator product AB:

$$\left(\hat{A} - \langle A\rangle\right) \cdot \left(\hat{B} - \langle B\rangle\right) \approx 0, \tag{4.53}$$

from which it follows that

$$\hat{A}\hat{B} \simeq \hat{A}\langle B\rangle + \langle A\rangle \hat{B} - \langle A\rangle\langle B\rangle. \tag{4.54}$$

We can apply this formulation to the scalar product of the spins to obtain

$$\hat{\boldsymbol{S}}_i \cdot \hat{\boldsymbol{S}}_j \simeq \hat{\boldsymbol{S}}_i \cdot \langle \boldsymbol{S}_j\rangle + \langle \boldsymbol{S}_j\rangle \cdot \hat{\boldsymbol{S}}_i - \langle \boldsymbol{S}_i\rangle \cdot \langle \boldsymbol{S}_j\rangle. \tag{4.55}$$

Therefore,

$$\hat{\mathcal{H}} = \frac{1}{2}\sum_{ij}{}' J_{ij}\, \hat{\boldsymbol{S}}_i \cdot \hat{\boldsymbol{S}}_j \simeq -\sum_{ij} J_{ij}\, \hat{\boldsymbol{S}}_i \cdot \langle \boldsymbol{S}_j\rangle - \frac{1}{2}\sum_{ij}{}' J_{ij}\, \langle \boldsymbol{S}_i\rangle \cdot \langle \boldsymbol{S}_j\rangle. \tag{4.56}$$

This form reduces to Eq. (4.36) when the total energy $\langle \mathcal{H}\rangle$ is calculated because the second term cancels the factor 2 in the first. Since at nonzero temperature the total energy represents the internal energy, which in turn fully characterizes the equilibrium state of the spin system, Eq. (4.40) should be equivalent to Eq. (4.36). Also note that in general

$$\begin{aligned}\hat{\mathcal{H}} = & -\sum_{ij}{}' J_{ij}\, \hat{\boldsymbol{S}}_i \cdot \langle \boldsymbol{S}_j\rangle - \frac{1}{2}\sum_{ij} J_{ij} \left(\hat{\boldsymbol{S}}_i - \langle \boldsymbol{S}_i\rangle\right) \cdot \left(\hat{\boldsymbol{S}}_j - \langle \boldsymbol{S}_j\rangle\right) \\ & + \frac{1}{2}\sum_{ij}{}' J_{ij} \langle \boldsymbol{S}_i\rangle \cdot \langle \boldsymbol{S}_j\rangle. \end{aligned} \tag{4.57}$$

We see that now the factor $\left(\frac{1}{2}\right)$ in the first (mean-field) term disappears (this term is partly compensated by the last term). Such a decomposition into mean-field and fluctuation parts reflects the appearance of a spontaneously broken symmetry state, and of inherent fluctuations around that state (the second term).

Below we treat only the classical fluctuations. For that purpose, we determine the free energy of the system via the partition function, which in turn requires evaluation of the trace of $\exp(-\beta\mathcal{H})$. When F is determined from

Eqs. (4.56) and (4.39), for a ferromagnet, we obtain the functional

$$F = -Nk_B T \ln \frac{\sinh\left[x\left(S+\frac{1}{2}\right)\right]}{\sinh\left(\frac{1}{2}\right)} + \frac{1}{2}\sum_{i\neq j} J_{ij}\left\langle S_i^z\right\rangle\left\langle S_j^z\right\rangle. \tag{4.58}$$

Formally, F is a functional of both H_a and magnetization M; hence, it represents a generalized form of the free energy as a function of state. The first term is associated with the mean-field dynamics, whereas the second cancels out the double counting of each interaction when taking the average over the spin degrees of freedom. Expanding the second term as before (see (4.46)), for cubic systems, we obtain

$$\frac{1}{2}\sum_{j(i)} J_{ij}\langle S_i^z\rangle\langle S_J^z\rangle \approx \frac{1}{2}J_0\left\langle S^z(\boldsymbol{r})\right\rangle^2 + \frac{1}{2}J_0 a_0^2\left\langle S^z(\boldsymbol{r})\right\rangle\nabla^2\langle S^z(\boldsymbol{r})\rangle\Big|_{\boldsymbol{r}=\mathbf{R}_i}. \tag{4.59}$$

We assume additionally that as we approach the phase-transition point $T = \Theta$, $\langle S^z(\boldsymbol{r})\rangle \ll S$, so that we may expand the hyperbolic function as

$$\sinh(ax) = ax + \frac{(ax)^3}{3!} + \frac{(ax)^5}{5!} + o(x^7). \tag{4.60}$$

In effect,

$$\begin{aligned}\frac{\sinh\left[\left(S+\frac{1}{2}\right)x\right]}{\sinh\left(\frac{x}{2}\right)} &\approx (2S+1)\frac{1+\left(S+\frac{1}{2}\right)^2\frac{x^2}{6}+\left(S+\frac{1}{4}\right)^4\frac{x^4}{120}}{1+\frac{x^2}{6}+\frac{x^4}{120}}+\cdots\\ &\approx (2S+1)\left[1+\frac{1}{6}S(S+1)x^2+\frac{1}{360}S(-1+2S+6S^2+3S^3)x^4+\cdots\right].\end{aligned} \tag{4.61}$$

Hence, the free energy per site can be written (to second order in x and with $H_a = 0$)

$$\begin{aligned}\frac{F}{N} \simeq{}& -k_B T\ln(2S+1) - \frac{S(S+1)}{6k_B T}\left[J_0\langle S^z(\boldsymbol{r})\rangle + J_0 a_0^2\,\nabla^2\langle S^z(\boldsymbol{r})\rangle\Big|_{\boldsymbol{r}=\mathbf{R}_i}\right]^2\\ &+\frac{1}{2}J_0\langle S^z\rangle^2 + \frac{1}{2}J_0 a_0^2\langle S^z(\boldsymbol{r})\rangle\nabla^2\langle S^z(\boldsymbol{r})\rangle\Big|_{\boldsymbol{r}=\boldsymbol{R}_i} + o\left(x^4\right).\end{aligned} \tag{4.62}$$

The first term is the free energy F_0/N of one noninteracting spin. Regrouping the terms, we obtain

$$\frac{F}{N} \simeq \frac{F_0}{N} + \frac{J_0}{2}\left(1 - \frac{\Theta}{T}\right)\langle S^z\rangle^2 - J_0\left(\frac{\Theta}{T} - \frac{1}{2}\right) a_0^2 \langle S^z\rangle \nabla^2 \langle S^z\rangle + o\left(x^4\right). \tag{4.63}$$

In the continuous-medium approximation F/N represents a free energy density. Therefore, the total free-energy functional in a space of d dimensions is

$$\begin{aligned} F = \int \frac{d^d x}{\Omega_0}\Bigg[\frac{F_0}{N} + \frac{J_0}{2}\left(1 - \frac{\Theta}{T}\right)\langle S^z\rangle^2 \\ - J_0\left(\frac{\Theta}{T} - \frac{1}{2}\right) a_0^2 \langle S^z\rangle \nabla^2 \langle S^z\rangle + o\left(x^4\right)\Bigg], \end{aligned} \tag{4.64}$$

where Ω_0 is the volume per spin (in the simple cubic case considered here, $\Omega_0 = a_0^3$). Integrating the third term by parts and neglecting the surface term (not always feasible), we obtain

$$\begin{aligned} F = F_0 + \int \frac{d^d x}{\Omega_0}\Bigg\{-\frac{J_0}{2}\left(\frac{\Theta - T}{T}\right)\langle S^z\rangle^2 \\ + J_0\, a_0^2\left(\frac{\Theta}{T} - \frac{1}{2}\right)\left[\nabla\langle S^z(\boldsymbol{r})\rangle\right]^2 + o\left(x^4\right)\Bigg\}. \end{aligned} \tag{4.65}$$

This expression coincides with the Landau expansion when $\langle S^z\rangle \equiv \langle S^z(\boldsymbol{r})\rangle$ is the order parameter and Θ, the mean-field critical temperature; (we have not evaluated explicitly the fourth-order term). Actually, in the Landau approach, we approximate T by Θ in the denominator of the first nontrivial term and in the second term, to obtain

$$\begin{aligned} F = F_0 + \int \frac{d^d x}{\Omega_0}\Bigg\{-\frac{3}{2S(S+1)} k_B(\Theta - T)\langle S^z\rangle^2 \\ + \frac{1}{2} J_0\, a_0^2\left[\nabla\langle S^z(\boldsymbol{r})\rangle\right]^2 + o\left(x^4\right)\Bigg\}. \end{aligned} \tag{4.66}$$

It is easy to prove that Eq. (4.52) is obtained variationally by taking F in the form (4.65) and applying to it the conditions for minimization of the functional:

$$\frac{\delta F}{\delta\langle S^z(\boldsymbol{r})\rangle} = 0, \qquad \frac{\delta^2 F}{\delta\langle S^z(\boldsymbol{r})\rangle^2} > 0. \tag{4.67}$$

This provides the equivalence between the mean-field approach and the Landau expansion as $T \to 0$. In particular, the Landau free-energy functional F reduces to the physical free energy if we substitute back into F the solution for $\langle S^z(\boldsymbol{r})\rangle$

obtained from condition (4.67). The Landau approach thus represents a *generalized thermodynamics*, in which the order parameter (magnetization) is included on equal footing with its conjugate parameters $\mathbf{H}_a$. Such generalized thermodynamics allows for a description of *continuous phase transitions* (see the next section) for a given lattice, and for a fixed range exchange interactions, in terms of the parameters that, in turn, determine both J_0 and Θ.

4.2.4 Landau Theory of the Ferromagnetic Transition

We will now concentrate on the relation of the ferromagnetic phase transition to the Landau approach. For this purpose, we must first introduce the higher-order term $\sim x^4$ in expansion (4.61) that we have so far neglected, and we also generalize the definition of x to include the applied magnetic field H_a. We return to Eq. (4.35) in the form $H = (g\mu_B)^{-1} J_0 \bar{S^z} + H_a$, $\bar{S^z} \equiv \sum_j \langle S_j^z \rangle$, and set $x = \beta(g\mu_B H_a + J_0 \bar{S^z})$. The lowest order term in expansion (4.61) will be truncated at the term linear in H_a; thus, $x^2 \approx \beta^2 (2g\mu_B J_0 \bar{S^z} H_a + J_0^2 \bar{S^z}^2)$, where the contribution $2\beta^2 g\mu_B J_0 \bar{S^z} H_a$ must now be added to (4.61). Additionally, we consider the fourth-order contribution in $x = \beta J_0 \bar{S^z}$. We then follow all of the steps that lead from (4.61) to (4.66) to obtain

$$\frac{F}{N} = \frac{F_0}{N} + \frac{3k_B(\Theta - T)}{2S(S+1)} \bar{S^z}^2 + \frac{1}{2} J_0 a_0^2 (\nabla \bar{S^z})^2 - \frac{1}{3} \frac{S(S+1)}{k_B T} g\mu_B J_0 \bar{S^z} H_a$$
$$+ \frac{\beta^3}{72} \left[S^2(S+1)^2 - \frac{S}{5} \left(-1 + 2S + 6S^2 + 3S^3 \right) \right] \left(J_0 \bar{S^z} \right)^4. \quad (4.68)$$

We next replace the coefficient of $(T - \Theta)\bar{S^z}^2$ by $\frac{A}{2}$, the coefficient of $\bar{S^z}^4$ by $\frac{B}{4}$, and the coefficient of $\left(\nabla \bar{S^z}\right)^2$ by C, and we set $\frac{S(S+1)J_0}{3k_B T} = \frac{\Theta}{T} \equiv \frac{T_c}{T} \approx 1$. This latter approximation is routinely used in correction terms for the temperature range near the transition. We thus obtain

$$\frac{\mathcal{F}}{N} = \frac{F_0}{N} - \frac{A}{2}(T_c - T)\bar{S^z}^2 + \frac{B}{4}\bar{S^z}^4 + C(\nabla \bar{S^z})^2 - g\mu_B \bar{S^z} H_a + o(\bar{S^z}^6), \quad (4.69)$$

where $A \equiv \frac{3k_B}{S(S+1)}$, $B \equiv \frac{\beta^2 J_0^4}{18}[\ldots]$, and $C \equiv \frac{1}{2J_0 a_0^2}$ are positive numbers. We have also replaced Θ with T_c, as we will discuss the magnetically ordered regime[7] $T < T_c$.

Consider first the zero-field spatially homogeneous solution, i.e., $\nabla \bar{S^z} = 0$. Under these conditions, the variational conditions (4.67) reduce to those for a

7. Obviously, in the saddle-point approximation, $T_c = \Theta$. This is not the case when fluctuations are taken into account.

minimum. Setting $\partial F/\partial \bar{S^z} = 0$ yields

$$\bar{S^z} = 0 \quad \text{or} \quad \bar{S^z} = \left[\frac{A}{B}(T_c - T)\right]^{\frac{1}{2}}. \tag{4.70}$$

The first (nonmagnetic) solution leads to the free energy $F = F_0$, whereas the second (magnetic) solution, which exists only for $T < T_c$, yields

$$\frac{F}{N} = \frac{F_0}{N} - \frac{A^2}{4B}(T_c - T)^2. \tag{4.71}$$

Thus the magnetic solution is physically stable for $T \leq T_c$. The mean-field critical exponent for magnetization is $\beta = \frac{1}{2}$ (since the critical exponent is defined through $\bar{S^z} \sim (T_c - T)^\beta$). The expression and results are the same as those discussed in Chapter 3 for the Ising model.

We can introduce a simple scaling of the free energy in the general case $\langle S^z \rangle = \langle S^z(\boldsymbol{r}) \rangle$ by defining the relative order parameter as

$$\eta(\boldsymbol{r}) = \frac{\bar{S^z}(\boldsymbol{r})}{\bar{S^z}} \equiv \frac{\bar{S^z}(\boldsymbol{r})}{\sqrt{\frac{A}{B}(T_c - T)}} \tag{4.72}$$

and by introducing an effective length $x \Rightarrow \frac{x}{\xi}$, where the characteristic distance (the correlation length) is given by

$$\xi \equiv \left[\frac{2BC}{A(T_c - T)}\right]^{-\frac{1}{2}}. \tag{4.73}$$

Under these conditions, the free energy expression requires an integration over all space:

$$\begin{aligned} F &= -Nk_BT\ln(2S+1) \\ &\quad - \frac{A^2}{4B}(T_c - T)^2 \int \frac{d^dx}{\Omega_0}\left\{2\eta^2(x) - \eta^4(x) + \left[\nabla\eta(x)\right]^2\right\}. \end{aligned} \tag{4.74}$$

The integration gives rise to a number. Hence,

$$F = F_0 - \alpha'\left(1 - \frac{T}{T_c}\right)^2. \tag{4.75}$$

In effect, the specific heat C_V (at constant volume) is of the form

$$C_V = -T\frac{\partial^2 F}{\partial T^2} = \frac{\alpha'}{2T_c^2}T. \tag{4.76}$$

Thus, C_V increases linearly with increasing temperature and jumps at $T = T_c$ by the amount $\Delta C_V = \frac{\alpha'}{2T_c}$. In real systems such as EuO, such a "lambda-type" behavior is observed; this is caused by the pronounced thermodynamic fluctuations near the critical point T_c.

Supplement: Zero-field Susceptibility in the Ferromagnetic Phase

We consider a ferromagnet subjected to a small applied magnetic field H_a. We now write the average z-component of the spin in the form $\bar{S^z} \equiv \bar{S^z} + \Delta\bar{S^z}$, where $\bar{S^z}$ is the component at $H_a = 0$, and $\Delta\bar{S^z}$ is the response to the field. In a spatially homogeneous state, we find from Eq. (4.69) that

$$-A\,(T_c - T)\,\bar{S^z} - A(T - T_c)\Delta\bar{S^z} + B\bar{S^z}^3 + 3B\bar{S^z}^2\Delta\bar{S^z} - g\mu_B H_a = 0. \quad (4.77)$$

Now we assume that the first and third terms sum to zero, as before, leaving

$$g\mu_B H_a = A\left[-(T_c - T) + \frac{3B\bar{S^z}^2}{A}\right]\Delta\bar{S^z}. \quad (4.78)$$

Introducing Eq. (4.70) and rewriting $A = 3k_B/S(S+1)$ in terms of C_M (Eq. (4.50)), we obtain

$$\Delta\bar{S^z} = \frac{g\mu_B H_a}{2A(T_c - T)} \quad (4.79)$$

and

$$M = g\mu_B\Delta\bar{S^z} = \frac{H_a(g\mu_B)^2}{2A(T_c - T)} = \frac{H_a C_M}{2(T_c - T)}, \quad (4.80)$$

whence the ferromagnetic susceptibility is given by

$$\chi = \frac{C_M}{2(T_c - T)}. \quad (4.81)$$

The susceptibility is of the same form for both the paramagnetic and for the ferromagnetic regimes ($\chi \sim \frac{C_M}{|T - T_c|}$), but in the ordered state χ is reduced by a factor of two. The diminution derives from the nonlinear part ($\sim B\bar{S^z}^4/4$) in the Landau expansion. However, this temperature evolution in a ferromagnet is usually obscured by the appearance of domain structures and by the irreversibility effects associated with it.

4.2.5 Mean-field Theory for Two-sublattice Antiferromagnets

Most systems that are *Mott–Hubbard insulators* are also antiferromagnets (i.e., have the opposite sign $J_{ij} \rightarrow -J_{ij}$ in Eq. (4.31)). Here we consider the simplest case of a two-sublattice antiferromagnet, which can exist in crystallographic structures that may be decomposed into two interpenetrating sublattices (examples are the *sc* or *bcc* lattices, but not the *fcc* lattice in three dimensions). Additionally, we assume that only one exchange integral (between A and B sublattices) is nonzero. In effect, we postulate a *Néel arrangement* of spins, i.e.,

$$\langle S_i^z \rangle = \begin{cases} \langle S_{Ai}^z \rangle = \bar{S^z} & \text{for} \quad i \in A \text{ sublattice}, \\ \langle S_{Bi}^z \rangle = -\bar{S^z} & \text{for} \quad i \in B \text{ sublattice}. \end{cases} \tag{4.82}$$

A sublattice with a given orientation of $\langle S_i^z \rangle$ is called *the Néel sublattice*. When $\langle S^z \rangle \neq 0$ in this case, both the time-inversion symmetry and the translational symmetry of the original lattice are spontaneously broken at the same time. Both symmetries are recovered above the phase transition temperature, *the Néel temperature* T_N.

To start formally, we introduce the effective fields acting on each sublattice (cf. Eq. (4.35)):

$$H_i^A \equiv -\frac{J}{g\mu_B} \sum_{j(i)} \langle S_{Bj}^z \rangle + H_a = \frac{J_0}{g\mu_B} \bar{S^z} + H_a, \tag{4.83}$$

$$H_j^B \equiv -\frac{J}{g\mu_B} \sum_{i(j)} \langle S_{Ai}^z \rangle + H_a = \frac{J_0}{g\mu_B} \bar{S^z} + H_a, \tag{4.84}$$

where we have set $J_{\langle ij \rangle} = -J$ and $J > 0$. The molecular fields on sublattices A and B differ in sign; hence we choose $\langle S_B^z \rangle = -\langle S_A^z \rangle$. The self-consistent equation for the order parameter is thus provided by Eq. (4.41). However, due to the change of sign of J_{ij}, we now write the free energy functional in the form

$$F = -Nk_B T \ln \frac{\sinh\left[x\left(S + \frac{1}{2}\right)\right]}{\sinh\left(\frac{x}{2}\right)} - \frac{1}{2} \sum_{i \neq j} J_{ij} \bar{S_i^z} \bar{S_j^z}. \tag{4.85}$$

Compared to (4.58), the last term has the opposite sign. However, since $S_{j(i)}^z = -S_i^z$ for the nearest neighbors, we recover the same expression as before.

To derive the static paramagnetic susceptibility in the limit $H_a \to 0$, we have to linearize the equation for the magnetization. We can also derive this result using more physical reasoning. Namely, the magnetic moment per atom on sublattice A is

$$M_A = g\mu_B \bar{S_A^z} = \chi_0 \left[-\frac{1}{g\mu_B} J_0 \bar{S_B^z} + H_a \right]. \tag{4.86}$$

In this equation, χ_0 is the magnetic susceptibility of the noninteracting spins ($\frac{C_M}{T}$ by the Curie law). The multiplier of χ_0 represents the total field acting on the A spins. Similarly,

$$M_B = g\mu_B \bar{S}^z_B = \chi_0 \left[-\frac{1}{g\mu_B} J_0 \, \bar{S}^z_A + H_a \right]. \tag{4.87}$$

Therefore, the two equations constitute a system of self-consistent relations for the expectation value of $\bar{S^z}$ in the applied field H_a:

$$\begin{cases} \bar{S}^z_A = \dfrac{\chi_0}{(g\mu_B)^2} \left(-J_0 \, \bar{S}^z_B + g\mu_B H_a \right), & \text{(a)} \\ \bar{S}^z_B = \dfrac{\chi_0}{(g\mu_B)^2} \left(-J_0 \, \bar{S}^z_A + g\mu_B H_a \right). & \text{(b)} \end{cases} \tag{4.88}$$

The total moment per site is $\frac{1}{2}(g\mu_B)^2(\bar{S}^z_A + \bar{S}^z_B)$ (note the factor $\frac{1}{2}$); hence, we obtain

$$\chi = \frac{g\mu_B(\bar{S}^z_A + \bar{S}^z_B)}{2H_a} = \frac{\chi_0}{1 + \frac{\chi_0 J_0}{(g\mu_B)^2}} \tag{4.89}$$

or explicitly, the expression

$$\chi = \frac{C_M}{T + \Theta_N}. \tag{4.90}$$

This has the same form as (4.49), except that now $\Theta \to \Theta_N$, where $\Theta_N = \frac{J_0 S(S+1)}{3k_B} = |\Theta|$. In general, the value of Θ_N is reduced by 10–30% from a true transition temperature T_N, $T_N < \Theta_N$. The reduction is due to the fact that in the mean field approach we do not take into account the destructive role of inhomogeneous thermodynamic fluctuations.

4.3 ROTATIONALLY INVARIANT FORM OF THE LANDAU FUNCTIONAL

So far we have studied the situation with one component order parameter $\eta(\boldsymbol{r})$. From the reasoning of Section 4.2.3 we see that we can repeat the whole procedure for each component of magnetization $M_\alpha(\boldsymbol{r}) \equiv \langle S^\alpha_i \rangle \big|_{\boldsymbol{i}=\boldsymbol{r}}$ with $\alpha = x, y, z$. This extension is particularly useful if we wish to include angular fluctuations of the magnetic moments. In that case the Landau energy density (4.69) assumes the additive form

$$\begin{aligned} \frac{\mathcal{F}}{V} = \frac{F_0}{V} - \frac{A}{2}(T_c - T) \sum_{\alpha=1}^{3} \bar{S^\alpha}^2 + \frac{B}{4} \sum_{\alpha=1}^{3} \sum_{\beta=1}^{3} \bar{S^\alpha}^2 \bar{S^\beta}^2 \\ + C \sum_{\alpha=1}^{3} \left(\nabla \bar{S^\alpha}\right)^2 - g\,\mu_B \bar{S^\alpha}\, \boldsymbol{H}_a + o((\bar{S^\alpha})^6). \end{aligned} \tag{4.91}$$

Now, introducing the vector order parameter $\boldsymbol{M}(\boldsymbol{r}) \equiv \left(\bar{S^x}(\boldsymbol{r}), \bar{S^y}(\boldsymbol{r}), \bar{S^z}(\boldsymbol{r})\right)$, we obtain the rotationally invariant form of the free energy

$$\frac{\mathcal{F}}{V} = \frac{F_0}{V} - \frac{A}{2}(T - T_c)\,\boldsymbol{M}(\boldsymbol{r})^2 + \frac{B}{4}\left[\boldsymbol{M}(\boldsymbol{r})^2\right]^2 + C\,(\nabla \boldsymbol{M}(\boldsymbol{r}))^2 \\ - g\mu_B \boldsymbol{M}(\boldsymbol{r}) \cdot \boldsymbol{H}_a + o(\boldsymbol{M}(\boldsymbol{r}))^6). \tag{4.92}$$

Obviously, the full Landau functional takes the form

$$\mathcal{F}\{\boldsymbol{M}(\boldsymbol{r})\} = F_0 + \int d^3r \left\{ \frac{A}{2}(T - T_c)\,\boldsymbol{M}(\boldsymbol{r})^2 + \frac{B}{4}\left[\boldsymbol{M}(\boldsymbol{r})^2\right]^2 \right. \\ \left. + C\,(\nabla \boldsymbol{M}(\boldsymbol{r}))^2 - g\mu_B \boldsymbol{M}(\boldsymbol{r}) \cdot \boldsymbol{H}_a \right\} + \cdots. \tag{4.93}$$

Note that now the equilibrium magnetization density is determined variationally from the condition

$$\frac{\partial \mathcal{F}}{\partial \boldsymbol{M}(\boldsymbol{r})} = 0, \tag{4.94}$$

which leads to the equation for the bulk magnetization

$$A(T - T_c)\boldsymbol{M}(\boldsymbol{r}) + B|\boldsymbol{M}(\boldsymbol{r})|^2\boldsymbol{M}(\boldsymbol{r}) - C\nabla^2\boldsymbol{M}(\boldsymbol{r}) - g\,\mu_B \boldsymbol{H}_a = 0, \tag{4.95}$$

which should be supplemented with proper boundary conditions. For an infinite system, $\nabla M_\alpha = 0$ as $|\boldsymbol{r}| \to \infty$.

4.4 OUTLOOK: MEANING OF MEAN FIELD THEORY

The role of this chapter is crucial for the further development of the theory of phase transitions. Therefore, we provide an overview of the main points of the Landau/mean-field approach:

1. The starting Landau functional (Landau Gibbs free energy) $\mathcal{L}$ depends in a functional manner simultaneously on both the order parameter η and on the canonically conjugated variable ($\boldsymbol{M}(\boldsymbol{r})$ and $\boldsymbol{H}_a$ in magnetic systems). This is not a contradiction, since we minimize the functional with respect to the former, and by substituting the minimal value into $\mathcal{L}$ we change $\mathcal{L}$ into the physical (Gibbs) free energy.
2. The Landau functional respects the full symmetry of the disordered (high-symmetry) state, but the solutions for $T < T_c$ that minimize it do not. In other words, the appearance of the nonzero value of the predefined order parameter is a necessary and sufficient condition to encounter a spontaneously broken symmetry state of the symmetry.

3. The singularities of the physically measurable quantities represent the essential singularities (not only the poles or finite discontinuities), as is the case in the Landau/mean-field theory. The critical behavior obtained within Landau theory coincides with that of mean field theory, since the latter reduces to the former when $T \to T_c$ (i.e., for $\eta \ll 1$).
4. The physics of continuous phase transitions basically involves singularities, both experimentally and theoretically. Obviously, we analyze the system in the neighborhood of the singular point.
5. Since we encounter singularities in the derivatives of the function of state $F(T, \boldsymbol{H}_a)$ (e.g., $C_p \sim \partial^2 F/\partial T^2$), there is no direct analytical continuation in mathematical sense from the regime $T > T_c$ to $T < T_c$. However, by defining the Landau functional we have extra variables, e.g., $\mathcal{L}(t, H_a; M)$, and we can encircle the singularity with a path that avoids the singularity. In effect, which makes it possible to specify a generalized function of state $F(T, H_a)$ over the full temperature range.
6. The basic question remains as to how to choose the broken symmetry state, say $+|\bar{\eta}|$, out of the two possibilities $\pm\bar{\eta}$. One way is to define spontaneous order in the limit as $H_a \to 0$ in a given direction. This is the approach of Bogoliubov, termed the method of *quasiaverages*. We will not concern ourselves with this problem since there are other dynamic factors to be discussed later, such as the *dynamical restoration of the full symmetry*, with the help of which the problem can be resolved. In any case, domains with magnetization $+\bar{\eta}$ and $-\bar{\eta}$ are experimentally observed in unprepared samples, such as in the absence of a field, which settles the issue from the experimental side.

Problem 4.1: A Simple Physical Example of a Continuous Phase Transition: Order–disorder Transformation

The metallic alloy CuZn contains exactly a fifty–fifty composition within the body centered cubic lattice (β-brass). At high temperatures ($T > T_c$) the A $\equiv$ Cu and B $\equiv$ Zn atoms are arranged in a random fashion on this *bcc* lattice (cf. Fig. 4.6A). At $T = 0$, the alloy is completely ordered, as shown schematically in Fig. 4.6B. The positions of A and B can be interchanged, but within a single domain of macroscopic size, only one of them gets selected. This is what is meant by spontaneous breakdown of discrete A$\leftrightarrow$B interchange symmetry.

The order parameter can be defined in a natural manner as follows: Suppose N_{AA} is the number of A atoms in A (corner) positions, whereas N_{AB} is those in B (center) positions. In the same manner, one can define N_{BB} and N_{BA}, respectively. We can define the degree of order as

$$\eta = \frac{N_{AA} - N_{AB}}{N_{AA} + N_{AB}} = \frac{N_{BB} - N_{BA}}{N_{BB} + N_{BA}} \tag{4.96}$$

For $N_{AB} = N_{BA} = 0$, we have complete order, $\eta = 1$. On the contrary, for $N_{AA} = N_{AB} = N_{BA} = N_{AB}$, we have $\eta = 0$, i.e., in this case, a continuous phase transition point. This situation is reminiscent of that for a binary mixture, which starts separating at $T = T_c$.

The question centers on the microscopic cause of the separation into ordered AB alloy, and why it takes the form of a phase transition. We recognize that the natural pair attraction $V_{AB} < 0$ must be stronger than the sum of V_{AA} and V_{BB}, where $i, j \equiv \langle i, j \rangle$, and $n_{iA,B}$ are the corresponding numbers $n_{iA,B} = 0, 1$ of atoms at site i. So the total energy of the system is

$$E = \sum_{\langle i,j \rangle} \left(V_{AB}\, n_{iA}\, n_{jB} + V_{AA}\, n_{iA}\, n_{jA} + V_{BB}\, n_{iB}\, n_{jB} \right). \tag{4.97}$$

Now, introducing the spin variables

$$n_{iA,B} = \frac{1}{2} + S_i^z, \tag{4.98}$$

where $S_i^z = \pm\frac{1}{2}$ for A or B atom occupancy, respectively, and similarly

$$n_{jB,A} = \frac{1}{2} - S_j^z, \tag{4.99}$$

we have, up to a constant (with the condition $n_A + n_B = 1$),

$$E = \sum_{\langle i,j \rangle} \left(-V_{AB}\, S_i^z\, S_j^z + V_{AA}\, S_i^z\, S_j^z + V_{BB}\, S_i^z\, S_j^z \right). \tag{4.100}$$

Hence, the system is described by the Ising model, with the effective exchange integral $J \equiv +(V_{AB} - V_{AA} - V_{BB})$, with an antiferromagnetic sign $J < 0$, unlike in Chapters 2 and 3, i.e.,

$$E = |J| \sum_{\langle i,j \rangle} S_i^z\, S_j^z. \tag{4.101}$$

Thus, the ordering in this spin representation is a checkerboard pattern, as shown in Fig. 2.2, and as can be shown explicitly, by subdividing the whole lattice of N sites into two sublattices, one with all spins up and the other with spins down. The solution of Eq. (4.101) is not simple; however, we can can generalize the mean-field solution to the case of two sublattices.

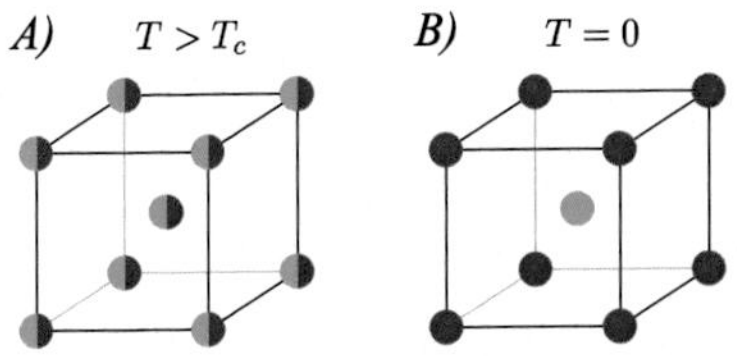

FIGURE 4.6 Schematic representation of the CuZn (AB) alloy on the bcc lattice in the disordered phase (A) and in the fully ordered state (B). For details, see the main text.

Important note

The separation into the two sublattices breaks not only the equivalence of occupancies between A and B, but simultaneously breaks the translational symmetry of the lattice: If the lattice parameter is a above T_c then it is $2a$ below T_c. We see that the phase transition may change the symmetry in the system in a nonobvious manner. See Fig. 4.6.

Problem 4.2: The Entropy and a Continuous Phase Transition for AB Alloys

So far we have proved that if $V_{AB} > V_{AA} + V_{BB}$, then the alloy will be completely ordered as shown in Fig. 4.6B. The question arises as to what is the critical temperature T_c for this order–disorder temperature. We consider the system at constant volume, and hence we can use the free energy $F = U - TS$ as a viable function of state. The entropy of that state is identical with that of an AB mixture (cf. Chapter 3 and Appendix 3.D in that chapter). Then the entropy is

$$S = \frac{N}{2}\left(n_A \ln n_A + n_B \ln n_B\right) \tag{4.102}$$

with $n_{A,B} \equiv \langle n_{A,B}\rangle \equiv \frac{1}{2} \pm \langle S_i^z\rangle$ and $N = N_A + N_B$, so that

$$S = \frac{N}{2}\left(\frac{1+2\langle S_A^z\rangle}{2}\ln\frac{1+2\langle S_A^z\rangle}{2} + \frac{1-2\langle S_B^z\rangle}{2}\ln\frac{1-2\langle S_B^z\rangle}{2}\right). \tag{4.103}$$

The Landau/Helmholtz energy in the mean field approximation is

$$\begin{aligned}\mathcal{F} = JzN\langle S_A^z\rangle\langle S_B^z\rangle - \frac{k_B T}{2}N\Bigg(&\frac{1+2\langle S_A^z\rangle}{2}\ln\frac{1+2\langle S_A^z\rangle}{2}\\ &+\frac{1-2\langle S_B^z\rangle}{2}\ln\frac{1-2\langle S_B^z\rangle}{2}\Bigg)\end{aligned} \tag{4.104}$$

with $z = 8$. Note, that we have used the standard expression for F and S, but now $\mathcal{F}$ is subject to the necessary condition

$$\frac{\partial \mathcal{F}}{\partial \langle S_A^z \rangle} = \frac{\partial \mathcal{F}}{\partial \langle S_B^z \rangle} = 0, \tag{4.105}$$

which leads to mean-field equations with $\langle S_A^z \rangle = -\langle S_B^z \rangle \equiv \bar{S}^z$ of the form (2.8) for $S = \frac{1}{2}$. The reader is encouraged to derive that equation in this alternative manner. The analysis is then the same as in Chapter 2.

Note on the entropy

If we define the order parameter for the sublattice A and B as $m \equiv 2|\langle S_{A,B}^z \rangle|$, then the entropy of the system can be written in the form

$$\frac{S}{N} = \frac{1}{2} k_B \left(\frac{1+m}{2} \ln \frac{1+m}{2} + \frac{1-m}{2} \ln \frac{1-m}{2} \right). \tag{4.106}$$

In the state with a perfect order, $m = 1$ and $S = 0$. On the contrary, for the completely disordered state, $m = 0$, and $S = Nk_B \ln 2$ assumes the maximal value. The latter state corresponds to the equal accessibility of each of the $\Omega = 2^N$ configurations. This provides an additional argument for the equivalence of the Ising system of fluctuating spin $S = \frac{1}{2}$ and the substitutionally disordered AB alloy. Both phase transitions belong to the same class of universality, about which we will write in detail later.

4.5 HISTORICAL NOTE: ORDER OF THE PHASE TRANSITION

Historically, phase transitions were divided by Ehrenfest (1935) into the first-, second-, and higher-order transitions.[8] The transition is of first order if the first derivatives of the function of state are discontinuous or singular. For example, in a magnetic system the quantities

$$\boldsymbol{M} = -\left.\frac{\partial F}{\partial \boldsymbol{H}_a}\right|_{T,H_a} \quad \text{and} \quad C_p = -\left.\frac{\partial U}{\partial T}\right|_{T,H_a=0} \tag{4.107}$$

change discontinuously at $T = T_c$. For second order transtions the first derivatives are continuous across T_c, but the second- and higher-order derivatives are discontinuous/singular at the critical point. In general, for the nth-order phase transition, the derivatives up to $(n-1)$th order are continuous across T_c, but starting from the nth order, they are either discontinuous or singular.

This division of phase transition according to their order is not universal since there exist phase transitions for which, e.g., the second derivative of F is singular, with respect to H_a, but not with respect to T (see e.g. as in spin glass

8. At equilibrium, there is no zeroth-order phase transition, since such a process would involve an energy change, which is inadmissible, and the system in equilibrium could then spontaneously radiate or absorb energy.

materials). Also, there are systems in which the transition is of infinite order. Those are the topological phase transitions in two dimensions (the Berezinskii–Kosterlitz–Thouless transition). In view of all this, L. Landau proposed that phase transitions be subdivided into just two groups: discontinuous and continuous transitions. The former can be called transitions of first order if all first derivatives of the function of state are discontinuous at the same critical point.

Chapter 5

More General Considerations Concerning Mean Field Theory: The Stratonovich–Hubbard Transformation*

*In Chapters 2–4, we have explored both the utility and the limitations of mean field theory (MFT). This leads, however, to a more general question: under what circumstances, if any, can we expect MFT to be useful? We now deal with this question; however, the discussion is only of indirect relevance to what is to follow; so the reader may wish to come back to this chapter, marked with *, after a first reading.*

5.1 GENERAL FORM OF THE PARTITION FUNCTION FOR THE ISING MODEL

Consider an Ising lattice of N spins in which every spin interacts equally with every other neighboring spin, so that the interaction energy per spin pair is $-\frac{2\mathcal{J}_T}{N}$, and let $N \to \infty$. The system is then so large that fluctuations may average to zero; MFT should then apply exactly. It seems reasonable that this same state could also be realized by assigning every spin an infinite number of nearest neighbors. We will now show that this is the case; this will lead us to study conditions under which MFT becomes applicable.

Assume that the Ising lattice is sufficiently large so that edge effects can be neglected. With $-\frac{\mathcal{J}}{N}$ as the pair energy, the Hamiltonian for these interactions becomes[1]

$$\mathcal{H} = -\frac{2\mathcal{J}_T}{N}\sum_{i<j} S_i S_j = -\frac{\mathcal{J}_T}{N}\sum_{i=1}^{N}\sum_{j=1}^{N} S_i S_j + \frac{\mathcal{J}_T}{N}\sum_{i=1}^{N} S_i^2. \tag{5.1}$$

On the right we have replaced the condition $i < j$ by an unrestricted double sum, so that every term with $i \neq j$ is counted twice; this is corrected by eliminating

1. Here, for simplicity, we assume that $S_i^z = \frac{1}{2}S_i$, where $S_i = \pm 1$. In effect, without loss of generality, we can define $\frac{\mathcal{J}_T}{4} \equiv J/4$ and proceed in the same manner as in the original Ising model.

A Primer to the Theory of Critical Phenomena. http://dx.doi.org/10.1016/B978-0-12-804685-2.00005-X

the factor of two. The indicated double sum also includes terms with $i = j$; these are eliminated by the last term on the right. Since $S_i^2 = 1$, the last term on the right is simply $\mathcal{J}_T$, so that

$$\mathcal{H} = \mathcal{J}_T - \mathcal{J}_T \left(\sum_{i=1}^{N} S_i\right)\left(\sum_{j=1}^{N} S_j\right) = \frac{\mathcal{J}_T}{4} - \frac{\mathcal{J}_T}{N}\left(\sum_{i=1}^{N} S_i\right)^2. \tag{5.2}$$

By introducing $\mathcal{K} \equiv \beta\mathcal{J}_T$ we may write the partition function in the form

$$\begin{aligned} \mathcal{Z}_N &= \sum_{\{S_i\}} \exp\left[-\mathcal{K} + \frac{\mathcal{K}}{N}\left(\sum_{i=1}^{N} S_i\right)^2\right] \\ &= \exp[-\mathcal{K}] \cdot \sum_{\{S_i\}} \exp\left[\left(\frac{\mathcal{K}}{N}\right)^{\frac{1}{2}}\left(\sum_{i=1}^{N} S_i\right)\right]^2. \end{aligned} \tag{5.3}$$

The determination of the partition function proceeds by use of the mathematical identity

$$\exp[q^2] = (2\pi)^{-\frac{1}{2}} \int_{-\infty}^{\infty} \exp\left[-\frac{x^2}{2} + 2^{\frac{1}{2}} q x\right] dx, \tag{5.4}$$

which is discussed in the Appendix.

By setting $q = \left(\frac{\mathcal{K}}{N}\right)^{\frac{1}{2}} \sum_{i=1}^{N} S_i$ we can rewrite the partition function in the format

$$\mathcal{Z}_N = \exp[-\mathcal{K}] \cdot \frac{1}{(2\pi)^{\frac{1}{2}}} \sum_{\{S_i\}} \int_{-\infty}^{\infty} \exp\left[-\frac{x^2}{2}\right] \cdot \exp\left[x\left(\frac{2\mathcal{K}}{N}\right)^{\frac{1}{2}} \sum_{i=1}^{N} S_i\right] dx. \tag{5.5}$$

The symbol $\sum_{\{S_i\}}$ is a shorthand notation for the summations $\sum_{S_1=-1}^{1} \cdots \sum_{S_N=-1}^{1}$ acting on the sum $\sum_{i=1}^{N} S_i$ in the exponential term. This representation of $\mathcal{Z}_N$ is called the *Stratonovich–Hubbard transformation*. It amounts to expressing the spin–spin (bilinear) interaction in the form that couples each individual spin to the effective field "x", which involves the Gaussian distribution. In this manner, the exact expression (5.5) generalizes the mean field approach, since the random field x has the Gaussian distribution. We obtain sums of ± 1 for each i in the second exponential term of (5.5) that are multiplied together N times, giving rise to

$$\mathcal{Z}_N = \frac{1}{(2\pi)^{\frac{1}{2}}} e^{-\mathcal{K}} \int_{-\infty}^{\infty} \exp\left[-\frac{x^2}{2}\right] \cdot \left[2\cosh x\left(\frac{2\mathcal{K}}{N}\right)^{\frac{1}{2}}\right]^N dx. \tag{5.6}$$

Now set $x = yN^{\frac{1}{2}}$, which converts the above to

$$\mathcal{Z}_N = \left(\frac{N}{2\pi}\right)^{\frac{1}{2}} 2^N e^{-\mathcal{K}} \int_{-\infty}^{\infty} \left[e^{-\frac{y^2}{2}} \cosh\left(y\sqrt{2\mathcal{K}}\right)\right]^N dy. \tag{5.7}$$

The partition function can be approximated in the lowest order by replacing the integral with its largest term. Since in the limit of large N the free energy per spin is specified by $F = -k_B T \ln[N^{-1}\mathcal{Z}_N]$, we seek the maximum of the function $f = -F/k_B T$, i.e.,

$$f = \ln\left[e^{-\frac{y^2}{2}} \cosh\left(y\sqrt{2\mathcal{K}}\right)\right] = -\frac{y^2}{2} + \ln\cosh\left(y\sqrt{2\mathcal{K}}\right). \tag{5.8}$$

The necessary condition for singling out the maximum reads $\frac{\partial f}{\partial y} = 0$. This leads directly to the relation

$$y = \sqrt{2\mathcal{K}} \tanh\left(y\sqrt{2\mathcal{K}}\right). \tag{5.9}$$

The functions y and f therefore involve solely the reduced interaction energy, relative to the thermal energy.

The above y optimizes the largest term that replaced the integral in Eq. (5.7), i.e., it thus determines the optimal value of the partition function (5.5) or (5.3). Equation (5.9) stands in the same relationship to Eq. (5.3) as Eq. (2.7) stands to Eq. (2.5) since (5.9) and (2.7) have the same functional form. Note that the first-order expansion of Eq. (5.9) yields $2\mathcal{K} = 2\mathcal{J}_T/k_B T_c$ or $\mathcal{J}_T = \frac{z}{2} J S^2$, where $\frac{zN}{2}$ is the number of bonds in an assembly of N members when each unit is surrounded by z nearest neighbors. This result is directly comparable to Eq. (2.11). In that sense the current derivation coincides with the molecular field equations in Chapter 2 or 3 that ignored all but nearest-neighbor interactions. As applied to the limit $N \to \infty$, the above objective of suppressing fluctuations is achieved by surrounding a given site by infinitely many neighbors. One way to achieve this scenario is to increase the number of dimensions d indefinitely, i.e., to take the limit as $d \to \infty$. This, for the first time, raises the issue of dimensionality; we can now ask whether we actually need to go to the infinity limit or whether a finite d value suffices for mean field theory to be viable. This problem will also be addressed in subsequent chapters.

5.2 THE UPPER CRITICAL DIMENSION

We assume that the spins are almost uniformly correlated over the so-called *correlation distance* ξ; the corresponding classical free energy density of a given fluctuation in a d-dimensional system is of order

$$\Delta F \sim \frac{k_B T}{\xi^d}. \tag{5.10}$$

Near T_c the correlation length conventionally varies with scaled temperature as $\xi \sim |t|^{-\nu}$, $t \equiv \frac{T-T_c}{T_c}$, whence

$$\Delta F \sim T_c |t|^{\nu d}. \tag{5.11}$$

We now use the heat capacity to estimate the *total* free energy F by use of the thermodynamic relation

$$C = -T\left(\frac{\partial^2 F}{\partial T^2}\right)_B = -\frac{T}{T_c^2}\left(\frac{\partial^2 F}{\partial t^2}\right)_B \sim |t|^{-\alpha}, \tag{5.12}$$

where the conventional divergence of C at the critical point was introduced on the right. Next, integrate twice with respect to t to obtain

$$F \sim T_c |t|^{2-\alpha}. \tag{5.13}$$

Thus, if we wish ΔF with $t \ll 1$ to be negligible compared with F as $t \to 0$, we must require that $\nu d > 2 - \alpha$. This relation is known as *Josephson's law* and will be properly derived later as an equality. Substituting the mean field values $\nu = \frac{1}{2}$ and $\alpha = 0$, we see that for MFT to be valid, we must require

$$d > d_{mf} = 4. \tag{5.14}$$

Here d_{mf} is called the *upper critical dimension*. Clearly, dimensionality and the special value $d = 4$ play an important role as we try to improve on MFT. The question still arises how large d must actually be before MFT becomes an acceptable approximation. For most purposes, $d = 4$ is adequate, a number that would seem not too different from our three-dimensional world. However, for $d < 4$, the situation is actually quite different, as we will soon see.

This completes our study of mean field approach.

5.3 APPENDIX: DERIVATION OF EQ. (5.4)

To derive Eq. (5.4), we start with the well-established relation (Poisson integral)

$$\int_{-\infty}^{\infty} dz\, e^{-z^2} = \sqrt{\pi} \tag{5.15}$$

and introduce the change of variable $z = \frac{x}{\sqrt{2}} - q$, whence

$$\frac{1}{\sqrt{2}}\int_{-\infty}^{\infty} dx \left[e^{-q^2} e^{\left(-\frac{x^2}{2}+\sqrt{2}xq\right)}\right] = \sqrt{\pi}, \tag{5.16}$$

or

$$e^{-q^2}\frac{1}{\sqrt{2\pi}}\int_{-\infty}^{\infty} dx \left[e^{\left(-\frac{x^2}{2}+\sqrt{2}xq\right)}\right] = 1, \tag{5.17}$$

from which Eq. (5.4) follows.

Exercise. Generalize transformation (5.4) to the Heisenberg interaction for classical spins.

Hint. Note that $\boldsymbol{S}_i \cdot \boldsymbol{S}_j = \sum_{\alpha=1}^{3} S_i^\alpha S_j^\alpha$ and decompose Eq. (5.4) into the product of three integrals. Note also that now in general all components fluctuate and that we now also deal with three random fields $\{x_i\}_{i=1,2,3}$.

Chapter 6

Generalities Relating to the Study of Critical Phenomena

Before we get into the study of critical phenomena beyond mean field theory, we survey the overall procedure on which the subsequent work is based. In linking observables to atomistic properties, one is faced with reducing about 10^{23} *microvariables to a few measurable parameters. In statistical thermodynamics, this is accomplished by the use of distribution functions that are approximated by their leading term, followed by the generation of a partition function that is then linked to thermodynamic variables of interest.*

6.1 INTRODUCTION

In the present context the approach is different: we are guided by the increasing size of the secondary phase as the system comes close to its critical point. In studying phase transitions, we therefore need to introduce a new characteristic length, the *correlation length* ξ that deals with the range of interactions among constitutive particles in a system. Consider again a lattice of spins in which at temperature $T = 0$ the magnetic moment M at one site tends to align neighboring spins along the same direction and vice versa, so as to reduce the energy of the system. This is reflected in an increase in $\xi \to \infty$. The opposite tendency is manifested at nonzero temperatures that generates spin disorder: entropy then "wins" when interacting spins become increasingly separated.

In Chapter 3, we formally introduced correlation functions as a means to quantify the above trend. In anticipation of later chapters, near a critical point, we encounter a correlation function of the form $\langle M(0)M(r)\rangle \sim e^{-r/\xi}$. However, precisely at T_c the correlation function displays an entirely different functional dependence: $\langle M(0)M(r)\rangle \sim r^{-q}$, which is an algebraic power law with $q > 0$. This structural difference carries an important message: Below T_c, under fixed conditions, the correlations are represented in terms of a characteristic length ξ that shows on average the distance range over which interactions predominate in forming fluctuations. At T_c the distance scale has lost its relevance as an indicator; the average is replaced by a range over which all lengths are of equal importance. In fact, as we have noted earlier, the fluctuations induced by correlations can and do extend over infinite distances at T_c. For $T < T_c$, statistic

A Primer to the Theory of Critical Phenomena. http://dx.doi.org/10.1016/B978-0-12-804685-2.00006-1

order extends over an infinite distance, whereas the fluctuations of this order do not.

The above reflects the fact that, under normal conditions, taking water as an example, we always encounter fluctuations in density on the scale of atomic distances. However, as we increase the temperature and pressure, the size and reach of these variations also increase; close to T_c, the fluctuations are comparable to the wavelengths of visible light, giving rise to the opaque appearance of the system. However, microscopic and small scale fluctuations are also still present, forcing us to deal with these many different length scales. This is where a scaling-up process comes in, which focuses on handling the large-scale properties, as we shall see. In this process, we shall also develop procedures for systematically reducing the large number of microscopic degrees of freedom.

We can raise the question how long-range correlations can develop from short-range interactions. Qualitatively, this takes place because of the feedback effect induced by the interactions between the neighbors, which results in establishing the total order, i.e., the broken-symmetry state of the whole system.

6.2 SCALING PROCEDURES: KADANOFF BLOCKS

There are two methods for handling the approach to the critical point and the concomitant increase in the characteristic length scale ξ. The first involves a scaling-up of events in direct space, based on a hypercubic lattice model of d dimensions. Here we follow a procedure pioneered by Kadanoff.[1] We begin at the atomic level and consider as the fundamental entities the spins occupying lattice sites separated at distances a along each cubic axis. We next subdivide the lattice into hypercubic blocks of edge lengths ba with integer $b > 1$, which are to be considered as new building blocks for a new lattice. All the spins within each block are then to be replaced by a single representative spin (the block spin), thereby diminishing the number of independent variables required to characterize the properties of the block lattice. As a last step, we diminish all dimensions of the newly generated spin lattice by the factor b; the new lattice then looks just like the original but contains fewer spins by a factor b^d. This process is then repeated, thereby increasingly winnowing out the number of degrees of freedom. Whereas we may not know the detailed mathematical structure of the lattices that are successively built up, we can compare the properties of the lattice before and after each blocking step. The above procedure is often referred to as a *coarse graining process*.

After every lattice-size increase, we examine the corresponding Hamiltonian. If properly constructed, after rescaling, the Hamiltonian will have the same

1. L. Kadanoff, *Physics* **2**, 263 (1966).

functional form as before, but with different coupling coefficients; details on how this is achieved will emerge later. In favorable cases, this process, known as the *renormalization group technique*, leads to a point where further scaling-up of the block size no longer changes the entire form of the Hamiltonian, signifying the presence of a critical point. Thus, as the winnowing proceeds, we hope to arrive at the point where a limiting Hamiltonian can be constructed with only a few parameters.

Rationale

The Kadanoff rescaling process reflects the fact that when we deal with vastly different length scales of fluctuations, a simple renormalization scheme, such as introducing electrons with an effective mass to simulate its interaction with the underlying lattice, no longer suffices. Instead, we need to adopt the repetitive coarse graining procedure that gradually eliminates the effects of short-range fluctuations, thereby focusing on the long-range interaction effects at the critical point. This then rationalizes the fact that completely different physical systems display the same critical phenomena. Due to the growing importance of the long-range phenomena, the atomic details become irrelevant. The resulting universality is described in terms of macroscopic concepts, such as the symmetry of the system, the range of interactions, and the dimensionality of the system.

The above process, when close to the critical point, is often visualized by referring to the examination of clusters in a sample through microscopes of different resolving power. Clusters of a given size appear to be smaller in the microscope of lesser resolution. We look at the same system, but under different magnifications. Very close to the critical point, we encounter an enormous range in sizes of fluctuating cluster islands, but averaged over time; the pictures revealed by both devices are very similar.

6.3 OPERATIONS IN RECIPROCAL SPACE

An alternative approach is to work in reciprocal space, which allows us ultimately to deal with continuous rather than discretized variables. This is not just a mathematical convenience: physical transitions occur as well in reciprocal space where fundamental constituents order around preferred momenta. As an example, we cite superfluid He, which is a normal liquid down to 2.17 K, below which it loses viscosity as the atoms gather collectively into a state of zero momentum. Superconductivity is another example, resulting from the pairing of electrons that produces boson-like entities with spin $S = 0$.

In working with momentum variables the basic idea is to systematically reduce the degrees of freedom of the wave number variables k needed to specify the properties of the system. The lattice at the atomic level is characterized by a lattice parameter a such that all wave numbers (momenta) with $|\boldsymbol{k}| \geq \Lambda \sim \pi/a$ can be ignored because we are not interested in what happens at subatomic scales; Λ serves as a cutoff parameter. We then select a set of wave numbers in the range $\Lambda/b \leq k \leq \Lambda(b > 1)$ and integrate their effects out in a manner detailed later; this corresponds to choosing a larger block size in the direct space. We then rescale the system in the wave number space, so as to remain comparable to the starting configuration. This step is repeated many times; such an approach generates systems with successively fewer wavelengths involved in constructing the approach to critical point. By use of the Fourier techniques we can ultimately transition to the continuum approach.

6.4 UTILITY OF ALGEBRAIC POWER LAWS

In deploying such a scaling process, we have altered the representation, but obviously not the physical properties, of the system. We therefore need a descriptor that reflects this coarse graining scale change while retaining the fundamental concept of invariance of physical properties. This is achieved by the use of algebraic power laws as we now show.

As a trivial example, consider the function $Q(t) = At^l$. Introduce $\tau = Rb^{\frac{q}{l}}t$ as a new variable that scales t as shown. The resultant $Q(\tau) = AR^{-l}b^{-q}\tau^l \equiv B\tau^l$ preserves the original functional form while altering the multiplying factor. In that sense the algebraic power law format of Q is insensitive to scale changes.

At a more sophisticated level, consider the *pair correlation function* of the general form that we deal with at a later stage

$$G(r) = \langle\phi(0)\phi(r)\rangle = \frac{D}{r^{d-2+\eta}}, \tag{6.1}$$

where $\phi(r)$ is a position-dependent *order parameter*, and where D is a constant. Now introduce the new distance scale $\rho = br$ and rescale by the scaling factor b^ω, whereby we generate the pair correlation function

$$G'(br) = b^{2\omega}\langle\phi(0)\phi(br)\rangle = \frac{Db^{2\omega}}{b^{-(d-2+\eta)}\rho^{d-2+\eta}}, \tag{6.2}$$

whereby we have replaced D by $Db^{2\omega}b^{d-2+\eta}$ but left the position dependence intact. To prove that the two versions are equivalent, set $2\omega = -(d-2+\eta)$, which reduces $G'(br)$ to its original $G(r)$ form. This is simply a more interesting case of a rescaling process that leaves the original functional form unaltered. By contrast, such a scale invariance does not hold for other types of functions, such as exponentials. Ultimately, however, we are going to have to rely on more than heuristic arguments to justify the use of power laws.

The above calls for a remark about the exponent in Eq. (6.1): As is evident from our earlier and later work, all mean field theories lead to correlation functions in which η vanishes. The more sophisticated renormalization theories described later show that η is of order 0.035. This at first glance may seem like a minor, unimportant departure from classical analysis. Not so! Once nonintegral power laws are encountered, the ordinary procedures of Newton and Leibniz for differentiating such functions no longer yield Taylor series expansions with positive integral powers. The $d-2$ portion of the exponent is termed the *canonical dimension*, and η is the *dimensional anomaly*. As proved later, the case $\eta = 0$ implies the neglect of fluctuations; the small parameter $\eta > 0$ is therefore a quantity of major importance that we must deal with in due time.

At this point, it should be mentioned that power laws are far more ubiquitous in science than their use in critical phenomena suggests. A large class of nonequilibrium phenomena falls into this category. Some of these spontaneously evolve into distinct classes of critical states without any external prodding, a property that comes under the designation of *self-organized criticality*. Phenomena in the category that display power law distributions include: the Pareto distribution of wealth in our society, the number of earthquakes releasing energies within a certain range, the distribution of towns with a specified population range, the distribution of words of different lengths in a language; or: the occurrence of volcanic eruptions, forest fires, traffic congestion, scientific citations, etc. Whereas their distributions are specified by a power law, none of these can be characterized by an average value because they are scale invariant. For a given process, the underlying mechanism is the same whether its manifestation involves large or small changes; huge earthquakes have enormous consequences relative to small tremors but fundamentally are not exceptional.

6.5 HOMOGENEOUS FUNCTIONS

We conclude with a brief foray into the concept of *homogeneous functions*. A function f of a single variable is homogeneous in degree n if $f(\lambda x) = \lambda^n f(x)$ for all λ. This feature can be extended to any number of independent variables: *Generalized homogeneous functions* of degree n satisfy the relation

$$f(\lambda^r x_1, \lambda^s x_2, \ldots) = \lambda^n f(x_1, x_2, \ldots) \tag{6.3}$$

for all λ. The utility of such functions in the development of the rescaling process will soon become evident. The important point is that functions other than power laws do not display this property and are therefore not suitable for scaling-up processes.

An immediate consequence of the above emerges for the particular case where we deal with two independent variables in the form

$$f(\lambda x, \lambda y) = \lambda^p f(x, y). \tag{6.4}$$

By adopting the choice $\lambda = 1/x$ we find that

$$f\left(1, \frac{y}{x}\right) = \left(\frac{1}{x}\right)^p f(x, y) \tag{6.5}$$

or

$$f(x, y) = x^p f\left(1, \frac{y}{x}\right); \quad \frac{f(x, y)}{x^p} = g\left(\frac{y}{x}\right). \tag{6.6}$$

Note how the two independent variables show up as a single ratio on the right.

How the homogeneity concept or scale invariance is used in dealing with critical phenomena via the elimination of microvariables will emerge as we go along. However, at this point, knowing nothing more, we can already study properties of the critical exponents introduced in Chapter 3, as we show below.

6.6 FIXED POINTS

As a final point of departure, we introduce the idea of a fixed point by a particular illustration: We follow the discussion by Gitterman and Halpern.[2] Consider the following mapping operation, called the *logistic map*:

$$x_{n+1} = s x_n (1 - x_n) \equiv f(x_n), 0 < x < 1, s > 1, \tag{6.7}$$

which provides a rule for relating two adjacent members of a sequence for the variable x. Fig. 6.1 shows the progression as n is increased. The straight line and parabola represent x_{n+1} and $f(x_n)$, respectively. Begin with a specific value for $x_0 > 0$ on the diagonal and increase it until it hits the parabola, where Eq. (6.7) is satisfied with $n = 1$. Shift to the right and set $x_1 \approx 0.25$ as the departure point from the diagonal and move upward up to its intersection at $f(x_2)$. Then repeat the process. The progression, depicted in the figure, closes toward a fixed end point

$$x^* = 1 - \frac{1}{s} \approx 0.625, \quad \text{where} \quad x^* = s x^* (1 - x^*). \tag{6.8}$$

The other intersection point occurs at $x^* = 0$.

2. M. Gitterman and V. Halpern, *Phase Transitions: A Brief Account with Modern Applications* (World Scientific, Singapore, 2004), Chapter 5.

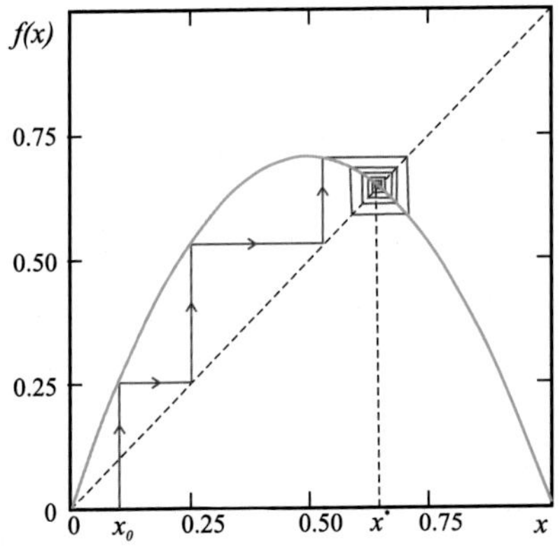

FIGURE 6.1 Plot illustrating the approach to a fixed point on a logistic map. The starting point is x_0, and by using the recurrence formula (6.7) we converge asymptotically to the fixed point x^*, which lies on the oblique dashed line $x = f(x)$. The consecutive values of x_n in each step $n > 1$ oscillate between the straight line and parabola shown in blue. For further explanation, see the main text.

The stability is studied by examining small excursions from x^*. Set $x_n = x^* + \delta_n$ and note that

$$\delta_{n+1} = \left(\frac{df}{dx}\right)_{x=x^*} \delta_n. \tag{6.9}$$

Then, if $\left|\left(\frac{df}{dx}\right)_{x=x^*}\right| < 1$, as is the case for $x = x^*$ since $\frac{df}{dx} = s(1-2x)$, the fixed point attracts other points in its immediate neighborhood. By contrast, at $x^* = 0$ the derivative exceeds unity and is repulsive. This is a neat illustrative example for what is to follow.

6.7 APPENDIX: A SIMPLE EXAMPLE

A matter frequently raised involves the question how a summation of analytic functions can give rise to a singular behavior. This delicate point can be usefully illustrated by considering a special example cited by Nishimori and Ortiz.[3] Consider the special function involved in the dynamic relation

$$\frac{du}{dt} = -2u\left(u^2 - 1\right), \tag{6.10}$$

3. H. Nishimori and G. Ortiz, *Elements of Phase Transitions and Critical Phenomena* (Oxford University Press, 2011), p. 56.

which has the solution

$$u(t) = \frac{u_0}{\left[u_0^2 - \left(u_0^2 - 1\right) e^{-4t}\right]^{1/2}}. \tag{6.11}$$

For any finite t, $u(t)$ is a continuous function of time; however, at infinite times, we get three (discontinuous) solutions: $u = 0, \pm 1$. On starting at $u_0 > 0$ or $u_0 < 0$, we end up at the fixed point $u^* = 1$ or $u^* = -1$, respectively. The point $u^* = 0$ is attained only for $u_0 = 0$ and is unstable.

Chapter 7

Failure of Mean Field Theory and Scaling Methods

The existence of correlated domains indicates the importance of units acting in concert and the concomitant irrelevance of local atomic properties of a given system. Thus, one looks for a methodology that takes account of the growth of correlated domains near the critical point, beginning with the atomic domain and ending up at the critical point, where the domain size reaches an infinite size. The general procedure relies extensively on the block method pioneered by Kadanoff[1] *that involves a coarse graining process. This method is best conveyed by once more resorting to a set of spins* $S_j = \pm 1$ *(in units of* $\frac{\hbar}{2}$*) distributed on sites* j *of a* d*-dimensional hypercubic lattice with lattice constant* a*.*

7.1 KADANOFF SCALING

As before, these spins interact with nearest-neighbor spins S_i via a common coupling constant $J > 0$ and with a superposed magnetic field B in the manner specified by the reduced Hamiltonian[2]

$$\beta\hat{\mathcal{H}} = -\beta J \sum_{\langle i,j\rangle} S_i S_j - \beta\mu B \sum_j S_j \equiv -K \sum_{\langle i,j\rangle} S_i S_j - H_m \sum_j S_j. \qquad (7.1)$$

The angular brackets denote summations over nearest-neighbor pairs; note the definition of the reduced parameters K and H_m (in units of $\beta \equiv 1/k_B T$).

Close to the critical point, islands of average "size" ξ, termed their *correlation* length, involve correlated spins that function more or less as a unit. It thus makes sense to introduce, by assembling the spins into groups, collective effects in adopting Kadanoff's procedure. Namely, we start with N unit cells of edge length a, whose ends bisect the distance between nearest neighbors, so that each unit cell contains one spin. We next subdivide the entire lattice into N' identical *blocks* of edge length ba, $b > 1$, an integer; each block then contains b^d original unit cells of edge length a. We characterize every such block by a

1. L. Kadanoff, *Physics* **2**, 263 (1966).
2. Note that taking the spin values $S = \pm 1$ is equivalent to taking $J_T = J/4$, which subsequently will be denoted by J. Likewise, μ is then $\mu/2 = \mu_B$.

A Primer to the Theory of Critical Phenomena. http://dx.doi.org/10.1016/B978-0-12-804685-2.00007-3

single associated spin S'_α ($\alpha = 1, 2, \ldots, N'$) that replaces the $\frac{N}{N'} = b^d$ original atomic spins S_j ($j = 1, 2, \ldots, N$). The choice of block spins is not unique: one procedure involves setting

$$S'_\alpha = \frac{1}{|\bar{M}_b| b^d} \sum_{j \in a} S_j \quad \text{and} \quad \bar{M}_b = \frac{1}{b^d} \sum_{j \in a} \langle S_j \rangle, \tag{7.2}$$

whereby each block spin has the same value $S'_\alpha = \pm 1$ as the atomic spins. By repeating this process we encompass increasingly larger domains.

Such coarse graining can have no effect on the physical properties of the lattice. If this is the case, we expect that *scale invariance* prevails. The various blocks generated in successive steps will have to have properties quite similar to the earlier ones, which encapsulates the essence of a *self-similarity* operation that is to be enforced during such coarse-graining. However, as already alluded to in Chapter 6, the coupling constants K and $\mathcal{H}_m$ do change at each iteration; these parameters are then said to *flow* in their parameter space. The process stops if a *fixed point* is reached, at which further coarse graining does not change the relevant parameters. The entire coarse graining procedure is termed the *renormalization group transformation* (RGT) for reasons that become evident later.

We now tentatively adopt an important assumption, namely, that the block spins S'_α are subject to the same interactions as the original spins S_j, so that Eq. (7.1) becomes applicable to the S'_α as well. This obviously requires a follow-up *rescaling*, whereby all linear distances, including the correlation lengths ξ, are reduced by the factor b: $\xi'_b = \frac{\xi}{b}$, $b > 1$. The lattice of spins S'_α for blocks of edge length ba before the rescaling is then restored to the same pattern as the original lattice of S_j spins for blocks of edge lengths a. Effectively, in shrinking the correlation lengths *after* every coarse graining step, we have actually moved the system further away from the critical temperature. Define the scaled temperature and scaled applied magnetic field in the following manner: $t \equiv \dfrac{T - T_c}{T_c}$ (briefly denoted by $\tilde{t}$ in Chapter 3), and $h \equiv \dfrac{H_m - H_{mc}}{H_{mc}}$.

Coarse graining, followed by rescaling to enforce constancy of the descriptive process, has thus diminished the scaled temperature from t to a new value, t_b. The interactions of S'_α with each other and with the magnetic fields have also been altered:

$$h \sum_j S_j \approxeq h\, \bar{M}_b\, b^d \sum_\alpha S'_\alpha \equiv h_b \sum_\alpha S'_\alpha, \quad h_b \equiv h\, \bar{M}_b\, b^d, \tag{7.3}$$

by which the scaled field has been renormalized.

Similarly, we need to rescale the interaction parameter so that K is replaced by K_b. The reduced Hamiltonian for the block spin is then specified by

$$\beta\hat{\mathcal{H}} \equiv -K_b \sum_{<\alpha,\beta>} S'_\alpha S'_\beta - h_b \sum_\alpha S'_\alpha. \tag{7.4}$$

The effective Hamiltonian for the block spins now has the same functional form as the original one. This then also applies to the partition functions and, by extension, to various thermodynamic functions, such as the chemical potential μ_c. Thus, we can render the Gibbs free energy $N'\mu_c\{S'_\alpha\}$ for the block spin assembly equivalent to that of the atomic spin assembly $N\mu_c\{S_j\}$.

Continuing this line of reasoning, the chemical potential, not being altered in this process, must be subject to the constraint that this rescaling involves a power law, as explained in Chapter 6, that is,

$$\mu_c(t_b, h_b) = b^d \mu_c(t, h). \tag{7.5}$$

We now need to specify the functional relation between t_b and t and between h_b and h. We tentatively invoke a second assumption, whose validity also remains to be investigated, namely, the interrelation

$$t_b = b^{y_t} t, \quad h_b = b^{y_h} h, \qquad y_t, y_h > 0, \tag{7.6}$$

which again rests on the previously mentioned assertion (Chapter 6) that power laws are appropriate in dealing with varying length scales. We then arrive at the scaling relation

$$\mu_c(b^{y_t} t, b^{y_h} h) = b^d \mu_c(t, h), \tag{7.7}$$

which may be rewritten in the standard form

$$\mu_c(\lambda^{a_t} t, \lambda^{a_h} h) = \lambda \mu_c(t, h), \tag{7.8}$$

where $a_t = \frac{y_t}{d}$ and $a_h = \frac{y_h}{d}$ are *scaling exponents* for the *scaling parameter*, all to be determined by experiment. What has been established so far by hand-waving arguments will be firmed up later on.

7.2 PROPERTIES OF THE HOMOGENEOUS EQUATION

To study the above relation, it is easiest to deal with a particular case: consider the Helmholtz free energy in the form (cf. Appendix A to Chapter 2)

$$dF = -SdT - V\mathcal{M}\,dH_m = N d\mu_c. \tag{7.9}$$

Now differentiate Eq. (7.8) with respect to h at constant t and V:

$$\lambda^{a_h} \frac{\partial \mu_c(\lambda^{a_t} t, \lambda^{a_h} h)}{\partial h} = \lambda \frac{\partial \mu_c(t,h)}{\partial h}, \tag{7.10}$$

which, in view of Eq. (7.9), produces a scaled equation that involves the magnetization $\mathcal{M}$,

$$\lambda^{a_h} \mathcal{M}(\lambda^{a_t} t, \lambda^{a_h} h) = \lambda \mathcal{M}(t,h). \tag{7.11}$$

This relation must hold for any λ of our choice; we find it advantageous to select $\lambda = \left(\frac{-1}{t}\right)^{\frac{1}{a_t}}$, where $t < 0$ below the critical point. We then find that

$$\mathcal{M}(t,h) = (-t)^{\frac{1-a_h}{a_t}} \mathcal{M}\left(-1, \frac{h}{(-t)^{\Delta}}\right), \quad \Delta \equiv \frac{a_h}{a_t} \quad (-t > 0), \tag{7.12}$$

showing that the magnetization close to the critical point involves a power law for the scaled temperature, and the dependence of $\mathcal{M}$ on a single variable, the ratio $\frac{h}{(-t)^{\Delta}}$; more on this below. Also, of interest is the quantity $\mathcal{M}$ under the condition $h \to 0$, so that

$$\mathcal{M}(t,0) = (-t)^{\frac{1-a_h}{a_t}} \mathcal{M}(-1,0), \tag{7.13}$$

which now involves solely a power law dependence on the scaled temperature. The factor on the right is a constant of no further interest.

The above may be compared to the experimental observation shown[3] in Table 2.1: As $t \to t_0^+$, $\mathcal{M} \sim (-t)^{\beta}$, whence we relate the empirical exponent β to the power law of Eq. (7.13):

$$\beta = \frac{1-a_h}{a_t}. \tag{7.14}$$

As the next step, following the above procedure, we adopt the choice $\lambda = h^{-\frac{1}{a_h}}$, whereby Eq. (7.11) assumes the form

$$\mathcal{M}(t,h) = h^{\frac{1-a_h}{a_h}} \mathcal{M}\left(1, \frac{t}{h^{\frac{1}{\Delta}}}\right), \tag{7.15}$$

which indicates that, close to criticality, the magnetization varies with the scaled magnetic field raised to some power, and that the function $\mathcal{M}$ depends on only

3. Conventional notation is introduced here and below, giving rise to an ambiguity. Be careful to distinguish between the t symbol introduced above and that used earlier, which relates to the reciprocal temperature.

a single variable, the ratio $\frac{t}{h^{\frac{1}{\Delta}}}$. Of interest is the special case $t = 0$, which leads to

$$\mathcal{M}(0,h) = h^{\frac{1-a_h}{a_h}} \mathcal{M}(1,0). \tag{7.16}$$

Again, we encounter a power law precisely at the critical temperature. For small fields, Table 2.1 indicates that the magnetic field dependence of the magnetization is conventionally expressed as $\mathcal{M} \sim h^{\frac{1}{\delta}}$, whence we establish the identification

$$\delta = \frac{a_h}{1-a_h}. \tag{7.17}$$

Equations (7.14) and (7.17) may now be solved for

$$a_t = \frac{1}{\beta(1+\delta)} \quad \text{and} \quad a_h = \frac{\delta}{1+\delta}, \tag{7.18}$$

thereby linking the critical exponents to experimental parameters. For future reference, note that

$$\Delta \equiv \frac{a_h}{a_t} = \beta\delta. \tag{7.19}$$

7.3 SCALING LAWS FOR SECOND DERIVATIVES

We now carry out the second differentiation of Eq. (7.11) with respect to h to determine the magnetic susceptibility χ. This leads to

$$\lambda^{2a_h}\chi(\lambda^{a_t}t, \lambda^{a_h}h) = \lambda\chi(t,h). \tag{7.20}$$

Here it is expedient to set $\lambda = (-t)^{-\frac{1}{a_t}}$, from which it follows that

$$\chi(t,h) = (-t)^{-\frac{2a_h-1}{a_t}}\chi\left(-1, \frac{h}{(-t)^{\Delta}}\right), \tag{7.21}$$

which, in the limit of a very small field magnetic field, reduces to the form

$$\chi(t,0) = (-t)^{-\frac{2a_h-1}{a_t}}\chi(-1,0). \tag{7.22}$$

This formulation may be compared to the empirical relation $\sim |t|^{-\gamma',-\gamma}$ for $T < T_c$ and $T > T_c$, respectively. We then see that the divergences in χ for $0 < a_h < \frac{1}{2}$, $a_t > 0$ are handled through the parameterization

$$\gamma' = \gamma = \frac{2a_h-1}{a_t}. \tag{7.23}$$

We can go further: let us insert (7.18) into the above. This leads to an interesting interrelation between critical exponents,

$$\gamma = \beta(\delta - 1), \tag{7.24}$$

which was first derived by Widom.[4] This relation has been extensively tested experimentally and found to be applicable to a wide variety of cases.

We may return to the function specifying $F(T, \mathcal{H}_m)$ and differentiate twice with respect to T to obtain $-\frac{C_h}{T}$, which involves the heat capacity at a constant (scaled) magnetic field. Translating to the scaled variable equation, we find that

$$\lambda^{2a_t} C_h(\lambda^{a_t} t, \lambda^{a_h} h) = \lambda C_h(t, h). \tag{7.25}$$

Now select $\lambda = (-t)^{-\frac{1}{a_t}}$. Then

$$C_h(t, h) = (-t)^{\frac{1-2a_t}{a_t}} C_h\left(-1, \frac{h}{(-t)^{\Delta}}\right), \tag{7.26}$$

which, in the absence of a field, reduces to

$$C_h(t, 0) = (-t)^{\frac{1-2a_t}{a_t}} C_h(-1, 0). \tag{7.27}$$

The variation of heat capacity with deviations from the critical temperature proceeds as conventionally specified by $C_h \sim (-t)^{-\alpha}$. This immediately establishes the correspondence

$$\alpha = -\frac{1 - 2a_t}{a_t}. \tag{7.28}$$

When substituting from Eq. (7.18), we obtain the interrelation

$$\alpha + \beta(1 + \delta) = 2, \tag{7.29}$$

which is known as Rushbrooke's relation.[5]

By the simple expedient of rewriting Eq. (7.29) as

$$\alpha + 2\beta\delta + \beta - \beta\delta = 2, \tag{7.30}$$

and then introducing Eq. (7.24), we obtain

$$\alpha + 2\beta\delta - \gamma = 2, \tag{7.31}$$

which is Griffith's scaling law.[6]

4. B. Widom, *J. Chem. Phys.* **43**, 3898 (1965).
5. G.S. Rushbrooke, *J. Chem. Phys.* **39**, 842 (1963).
6. R.B. Griffiths, *Phys. Rev. Lett.* **14**, 623 (1965).

7.4 SUMMARY AND REMARKS

In summary, we have illustrated how the coarse-graining method and the use of homogeneous functions lead to the identification of critical exponents in terms of empirically measurable parameters and to the establishment of interrelations between them that might not have been foreseen. The present listing is not exhaustive; we later encounter additional connections between the critical exponents.

Equations (7.12), (7.15), (7.21), and (7.22) reveal a very interesting aspect of the homogeneity requirement alluded to earlier: As an example, consider Eq. (7.12) and take the absolute values of the scaled temperature, so as to deal with t values both above and below T_c. Using Eq. (7.19), this equation may then be rewritten as

$$\frac{\mathcal{M}(t,h)}{|t|^\beta} = \mathcal{M}\left(\frac{h}{|t|^{\beta\delta}}\right), \tag{7.32}$$

so that magnetization data plotted as shown on the left should be representable as a universal function of $\frac{h}{|t|^{\beta\delta}}$, with one branch above and one branch below T_c. We will leave it to the reader as an exercise to deal in similar fashion with the remaining equations. This then illustrates and explains the origin of the data collapse that we noted earlier. In fact, the collapse, such as illustrated in Fig. 7.1, serves as a warrant for the applicability of the above approach to critical phenomena. This live example also shows how one invokes alternative formulations of Eq. (7.10) (cf. the measurements by Green et al.[7]), who used published *PVT* studies on Xe, Ar, SF_6, N_2O, and $CClF_2$ near their respective liquid–gas critical points. Note that they have chosen to plot the data as shown in Fig. 7.1 on a log–log scale as $\frac{|\Delta\mu_c|}{|t|^{\beta\delta}}$ vs. $\frac{|\Delta\rho|}{|t|^\beta}$. Numerical integrations of the Gibbs–Duhem relation were used to convert the data to chemical potentials. The data collapse into a branch above and below the critical temperature is obvious. The theoretical basis as a grounding for this analysis is provided in Appendix A.

The above approach provides a way of transition from atomic to macroscopic properties without applying the traditional methodology of statistical mechanics that eliminates the atomic degrees of freedom in one go. The gradual thinning-out of these degrees of freedom by the Kadanoff procedure ultimately rests on the fact that the coupling constants and independent variables change predictably with alterations in the length scale.

7. M.S. Green, M. Vicentini-Missoni, J.M.H. Levelt Sengers, *Phys. Rev. Lett.* **18**, 1113 (1967).

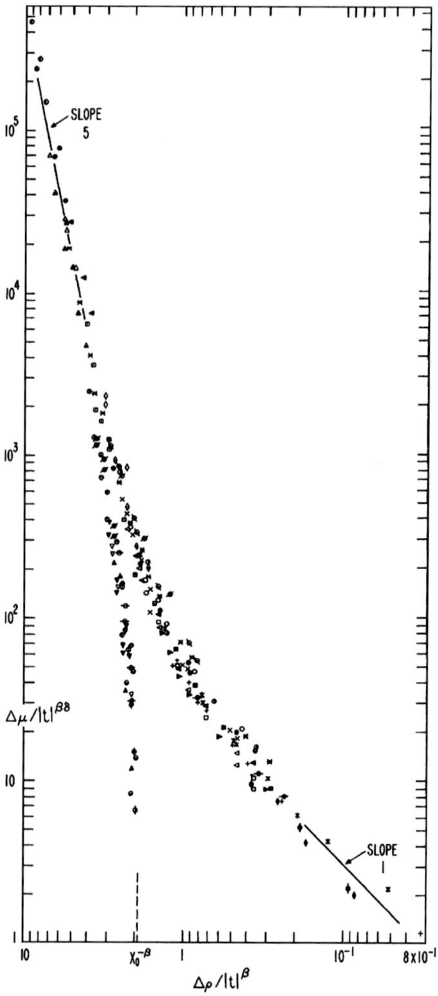

FIGURE 7.1 Plot on a log–log scale of the scaled chemical potential $\Delta\mu/|t|^{\beta\delta}$ versus the scaled density $\Delta\rho/|t|^{\beta}$ for a variety of liquid–gas systems near their critical points. Note the collapse of the data to two sets for $T \geqslant T_C$ and $T < T_C$, respectively, onto a branch above and below the critical temperature. The critical exponents are $\beta = 0.35$ and $\delta = 5$. For details, see Appendix A and Green et al. (see footnote 7).

7.5 APPENDIX A: SCALING LAW FOR THE CHEMICAL POTENTIAL

We assume that the chemical potential has the form $\mu_C = \mu_C(t, p)$, where t and p are the scaled temperature and pressure; v is defined similarly. Then for present purposes, Eq. (7.10) is rewritten in the form

$$\lambda^{a_p} d\mu_c(\lambda^{a_t} t, \lambda^{a_p} p) = \lambda d\mu_c(t, p). \tag{7.33}$$

Now introduce the Gibbs–Duhem relation, Eq. (3.C.4), and Eq. (3.46) to write

$$d\mu_c\Big|_t = v dp = \left(\frac{-2p}{3}\right)^{\frac{1}{\delta}} dp, \tag{7.34}$$

which may be integrated to yield

$$\Delta\mu_c = \left(\frac{-2}{3}\right)^{\frac{1}{\delta}} \left(p^{\frac{1}{\delta}}\right)^{1+\delta} = -\frac{3}{2} v \cdot v^{\delta} = v\left(2|t|^{\beta}\right)^{\delta} = 2^{\delta}\left(-\frac{3}{2}\right) v|t|^{\beta\delta}, \tag{7.35}$$

where Eq. (3.44) was introduced at the end. Now select the convenient form $\lambda = |t|^{\frac{-1}{a_t}} = |t|^{-\beta(1+\delta)}$, where Eq. (7.18) was invoked. Then,

$$\lambda\Delta\mu_c = 2^{\delta}\left(\frac{-3v}{2}\right)|t|^{-\beta}, \tag{7.36}$$

which attends to the right-hand side of Eq. (7.33). On adopting Eq. (7.19) the integrated form of Eq. (7.33) reads

$$\frac{\Delta\mu_c\left(1, \frac{\left(-\frac{3}{2}\right)v^{\delta}}{|t|^{\beta\delta}}\right)}{|t|^{\beta\delta}} = 2^{\delta}\left(-\frac{3}{2}\right)\frac{v(t, p)}{|t|^{\beta}}. \tag{7.37}$$

It remains to note that, close to the critical point, we may introduce, with no appreciable error, the density

$$v = \frac{V - V_c}{V_c} = \frac{\rho_c - \rho}{\rho} \approx -\frac{\rho - \rho_c}{\rho_c} \equiv -\Delta\rho \tag{7.38}$$

which leads directly to the data plots in Fig. 7.1.

Chapter 8

Kadanoff Scaling

The idea of Kadanoff scaling of the lattice is elaborated in detail for Ising spins.

8.1 EXAMPLE INVOLVING A RING

We now illustrate and expand the somewhat austere presentation of Chapter 7 to provide live examples illustrating the rescaling operations of functions. The discussion follows the presentation by Maris and Kadanoff[1] and McComb.[2] For ease of visualization, consider a circular ring of N (a very large number) of lattice points $i = 1, 2, \dots, N$, separated by a uniform distance a. Each site contains a magnetic dipole whose "spin" is represented by $S_i = \pm 1$, corresponding to the "spin-up" and "spin-down" configurations in the direction perpendicular to the ring. By way of review, in a uniform magnetic field B, every dipole acquires an energy $E_B = -\mu B$ or $E_B = \mu B$ in its "spin-up" or "spin-down" state. Each dipole interacts as well with its nearest-neighbor spin S_j at $j = i + 1$ and at $j = i - 1$, characterized by an interaction energy $J > 0$, which can be estimated via the exchange energy between spin pairs. Accordingly, the interaction energies are represented by $E_J = -J S_i S_j$, whether the spins are in parallel ($S_i S_j = 1$) or antiparallel ($S_i S_j = -1$) alignment. All longer-range magnetic interactions are ignored. In this primitive example, the total energy of the system is specified by ($\mu \equiv g\mu_B$)

$$E = E_B + E_J = -\mu B \sum_{i=1}^{N} S_i - J \sum_{\langle i,j \rangle} S_i S_j, \tag{8.1}$$

where the angular brackets indicate the restriction to nearest-neighbor locations. We must be careful, in the manner shown further, to count each i, j interaction only once.

The connection to macroscopic properties involves the determination of the partition function, summed over all states $\{S_i\}$ consisting of the configurations

1. H.J. Maris and L. Kadanoff, *Am. J. Phys.* **46**, 652 (1978).
2. W.D. McComb, *Renormalization Methods* (Oxford University Press, 2004), Sec. 2.1.2.

A Primer to the Theory of Critical Phenomena. http://dx.doi.org/10.1016/B978-0-12-804685-2.00008-5

of the set of values $\{S_i\}$:

$$\mathcal{Z}_0\left(B, \frac{J}{k_{\mathcal{B}}T}\right) = \sum_s \exp\left\{-\frac{E_s}{k_{\mathcal{B}}T}\right\}$$
$$= \sum_{\{S_i\}} \exp\left\{\frac{\mu B \sum_{i=1}^{N} S_i + J \sum_{<i,j>} S_i S_j}{k_{\mathcal{B}}T}\right\}. \tag{8.2}$$

We now specialize to spontaneous magnetizations in the absence of an applied magnetic field, for which

$$\mathcal{Z}_0(K) = \sum_{\{S_i\}} \exp\left\{\frac{J}{k_{\mathcal{B}}T} \sum_{<i,j>} S_i S_j\right\} \equiv \sum_{\{S_i\}} \exp\left\{K \sum_{<i,j>} S_i S_j\right\}$$
$$= \sum_{\{S_i\}} \prod_{<i,j>} \exp\{K S_i S_j\}, \tag{8.3}$$

where we have again introduced the *reduced coupling constant* $K \equiv \frac{J}{k_{\mathcal{B}}T}$; we also converted the exponent of a sum of terms into a product of exponential factors, each of which deals with the energetics of nearest-neighbor interactions. We render this restriction explicit by rewriting the partition function in groups of two terms as

$$\mathcal{Z}_0(K) = \sum_{\{S_i\}} \prod_i \exp\{K S_i S_{i+1}\} = \sum_{\{S_i\}} e^{K(S_1S_2+S_2S_3)} e^{K(S_3S_4+S_4S_5)} \cdots. \tag{8.4}$$

Now insert $S_2 = \pm 1$, which appears only in the first factor; then

$$\mathcal{Z}_0(K) = \sum_{\{S_{i\neq 2}\}} \left[e^{K(S_1+S_3)} + e^{-K(S_1+S_3)}\right] e^{K(S_3S_4+S_4S_5)} \cdots. \tag{8.5}$$

In a similar fashion, set $S_4, S_6, \ldots = \pm 1$ and continue the summation process, whereby we convert the original partition function to the form

$$\mathcal{Z}_1(K) = \sum_{\{S_{2i-1}\}} \left[e^{K(S_1+S_3)} + e^{-K(S_1+S_3)}\right]\left[e^{K(S_3+S_5)} + e^{-K(S_3+S_5)}\right] \cdots. \tag{8.6}$$

The partition function now omits all even-numbered spins. In effect, we have created a partition function for a ring that skips all sites originally indexed with even labels, which is a representative example of the Kadanoff *coarse graining* procedure.

We now ask the question: Is it possible to render Eq. (8.6) in a format similar to (8.3), namely,

$$\mathcal{Z}_1(K') = \sum_{\{S_j\}} \prod_j f(K) \exp\left\{K' S_j S_{j+1}\right\}, \tag{8.7}$$

where the index j runs only over all odd-numbered spins, f is a function of the original K, and where K' can be reexpressed solely in terms of K? If so, then all thermodynamic functions that are based on the use of $\ln \mathcal{Z}_0$ or $\ln \mathcal{Z}_1$ and on the temperature derivatives differ at most by an extra term involving the function $f(K)$.

What have we so far accomplished? Essentially, we have related the thermodynamics of the original ring to that in which half of its members are missing. If this ring with separation distances $2a$ between constituents is now scaled back so that all odd numbered members are separated by the distance a, then the ring has been reduced to its original size, with its topology intact; both physically and mathematically, we have achieved scale invariance, our original goal.

To determine the relation between $\mathcal{Z}_0(K)$ and $\mathcal{Z}_1(K')$, we examine a typical term in the proposed interrelation:

$$\exp\{K(S_i + S_{i+1})\} + \exp\{-K(S_i + S_{i+1})\} = f(K) \exp\left\{K' S_i S_{i+1}\right\}. \tag{8.8}$$

There are two cases to consider: (a) The spins at i and $i+1$ are in parallel orientation; then $S_i = S_{i+1} = 1$ or -1. In either case, we find that

$$\exp(2K) + \exp(-2K) = f(K) \exp(K'), \tag{8.9}$$

which we rewrite as

$$f(K) e^{K'} = 2\cosh(2K). \tag{8.10}$$

(b) The two spins are in antiparallel orientation; then $S_i = -S_{i+1} = 1$ or -1. In either case, we find that

$$f(K) e^{-K'} = 2. \tag{8.11}$$

Now eliminate $f(K)$ in Eqs. (8.10) and (8.11):

$$e^{2K'} = \cosh(2K), \tag{8.12}$$

whence

$$K' = \frac{1}{2} \ln \cosh(2K), \tag{8.13}$$

which achieves our goal of specifying K' in terms of K. Next, take the square root of (8.12) and substitute this into (8.11) to obtain

$$f(K) = 2\cosh^{\frac{1}{2}}(2K), \tag{8.14}$$

which solves the problem at hand. On comparing Eq. (8.7) with Eq. (8.3), we can relate the two partition functions in the following symbolic form:

$$\mathcal{Z}(N, K) = f^{\frac{N}{2}}(K) \cdot \mathcal{Z}\left(\frac{N}{2}, K'\right). \tag{8.15}$$

The original partition function (PF) for N units has now been rewritten as a PF of the same form for $\frac{N}{2}$ units, with altered coupling constants (CCs). The PF for $\frac{N}{2}$ units is multiplied by a function involving the original CCs.

Since the free energy of a system changes linearly with the number of constituent particles, we note that $-\beta \ln Z = N\mu_z$, where μ_z, proportional to the chemical potential, is size-independent. Then, taking logarithms of Eq. (8.15), we find that

$$\mu_z(K) = \frac{1}{2}\ln f(K) + \frac{1}{2}\mu_z(K'), \tag{8.16}$$

or

$$\mu_z(K') = 2\mu_z(K) - \ln\left[2\cosh^{1/2}(2K)\right]. \tag{8.17}$$

Equation (8.17) is the essential result, which is later rederived by the so-called renormalization group analysis. Thus, if $\mathcal{Z}(N, K)$ is known for one value of the coupling constant K, then the above recursion relation is used to determine the partition function for other values K'. Once μ_z is known, all thermodynamic quantities of interest can be deduced. As written above, the "renormalization process" accompanying the above rescaling always lowers the coupling parameter: $K' < K$ at every repetition of the scaling process, a feature we had already encountered in Chapter 7.

We may also move in the opposite direction by solving Eq. (8.13) for

$$K = \frac{1}{2}\cosh^{-1}\left(e^{2K'}\right), \tag{8.18}$$

and, via Eq. (8.12),

$$\mu_z(K) = \frac{1}{2}\ln 2 + \frac{1}{2}K' + \frac{1}{2}\mu_z(K'). \tag{8.19}$$

Thus, repeated rescalings lead either to an unending growth of $K \to \infty$, or to its minimum positive value $K \to 0$. Both are known as *trivial critical points*. In the above processes, the coupling constant is said to "flow" with each iteration.

We do not encounter a stationary intermediate point, indicative of a nontrivial critical point.

8.2 THE CASE OF TWO DIMENSIONS

Let us now apply the above method to a two-dimensional square planar array of lattice sites depicted in Fig. 8.1. Neglecting edge effects, any lattice site is surrounded by and interacting with its four nearest neighbors $i = 1, \ldots, 4$, as indicated by the connecting bonds. Now omit half of the original sites, namely those marked by the crossed circles, to generate Fig. 8.2; then rotate the set of sites counterclockwise by 45° and shrink the lattice back to its original size, as required by the Kadanoff scaling process. The resulting nearest-neighbor interactions are indicated by the bonds that lie along the original diagonal directions. By analogy to Eq. (8.6) a typical term in the resulting partition function reads:

$$\begin{aligned} \mathcal{Z}_1(K) = \sum_{\{S\}} \ldots \exp\{K(S_1 + S_2 + S_3 + +S_4)\} \\ + \exp\{-K(S_1 + S_2 + S_3 + S_4)\} \ldots \end{aligned} \tag{8.20}$$

for every square in Fig. 8.2. In this process, we generate not only the usual nearest-neighbor interactions but also interactions among more distant neigh-

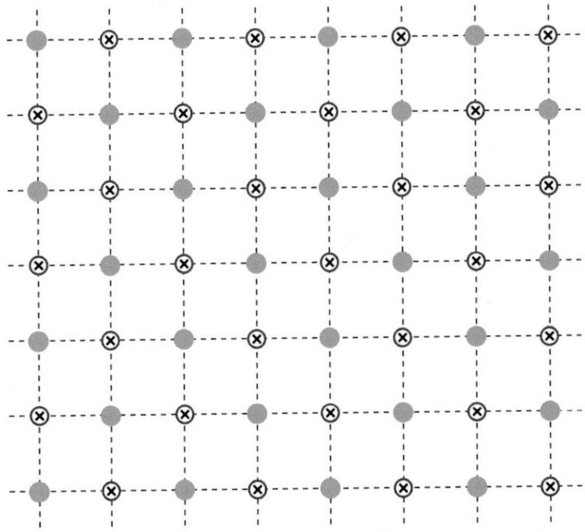

FIGURE 8.1 Portion of a two-dimensional Ising lattice of spins. Sites marked with a cross are marked for deletion during the Kadanoff rescaling. Spin–spin interactions are limited to nearest-neighbor sites.

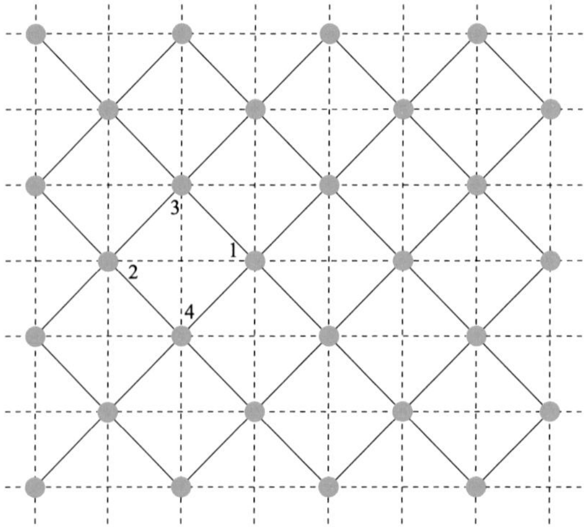

FIGURE 8.2 The two-dimensional square lattice after removal of half of the original sites. Nearest-neighbor interactions are indicated by the bonds. The lattice should be viewed after counterclockwise rotation by 45° and shrinking of unit cell length by $\sqrt{2}$.

bors. For a typical term, this means that we try to match

$$\begin{aligned} \exp\{K(S_1+S_2+S_3+S_4)\} &+ \exp\{-K(S_1+S_2+S_3+S_4)\} \\ &= f\cdot\exp\left\{\frac{1}{2}K_1(S_1S_2+S_2S_3+S_3S_4+S_4S_1)\right. \\ &\quad \left.+K_2(S_1S_3+S_2S_4)+K_3S_1S_2S_3S_4\right\}. \end{aligned} \tag{8.21}$$

If we now set $S_i=+1$ or $S=-1$ for all i, then we obtain

$$e^{4K}+e^{-4K}=fe^{2K_1+2K_2+K_3}. \tag{8.22}$$

Proceeding similarly with patience for all other possible $S_i=\pm1$ combinations that leave no net unmatched spin on the lattice, we find that

$$2=fe^{-2K_1+2K_2+K_3}, \tag{8.23}$$

$$e^{2K}+e^{-2K}=fe^{-K_3}, \tag{8.24}$$

$$2=fe^{-2K_2+K_3}. \tag{8.25}$$

You should verify that by introducing hyperbolic cosines and taking logarithms these equations may be solved for

$$K_1 = \frac{1}{4}\ln\cosh(4K),$$
$$K_2 = \frac{1}{8}\ln\cosh(4K), \tag{8.26}$$
$$K_3 = \frac{1}{8}\ln\cosh(4K) - \frac{1}{2}\ln\cosh(2K).$$

Thereby the expression analogous to Eq. (8.15) becomes (for a more detailed discussion, see Nishimori and Ortiz[3])

$$\mathcal{Z}(N, K)$$
$$= f^{\frac{N}{2}}(K) \cdot \sum_{\{S\}} \exp\left[K_1 \sum_{nn} S_r S_s + K_2 \sum_{nnn} S_r S_s + K_3 \sum_{squ} S_p S_q S_r S_s\right], \tag{8.27}$$

where the solution of Eqs. (8.22)–(8.25) leads to

$$f(K) = 2\cosh^{\frac{1}{2}}(2K)\cosh^{\frac{1}{8}}(4K). \tag{8.28}$$

Thus, while the various K_i have been expressed in terms of K and while $\mathcal{Z}(N, K)$ has been rescaled, we have not been able to achieve the sought-after simple form analogous to Eq. (8.15). Some ingenuity is now required to achieve this goal.

If we simply omit K_2 and K_3, then we find that $K' \equiv K_1$ is given by Eq. (8.26) and

$$\mu_z(K') = -\ln\left\{2\cosh^{\frac{1}{2}}(2K)\cosh^{\frac{1}{8}}(4K)\right\} + 2\mu_z(K), \tag{8.29}$$

which is analogous to what we encountered in the one-dimensional case; we have uncovered nothing new.

The next attempt depends on plausibility arguments. We drop the term involving K_3 and set

$$K' = K_1 + K_2 = \frac{3}{8}\ln\cosh(4K). \tag{8.30}$$

The function (8.29) remains unaltered. However, this adjustment produces an interesting new feature: numerical analysis shows that, for $K > K_c \approx 0.507$, successive rescalings increase K, whereas for $K < K_c$, successive rescalings decrease K. This will later be shown to be an indicator (necessary but not sufficient) for the existence of a nontrivial critical point. We have discovered

3. H. Nishimori and G. Ortiz, *Elements of Phase Transitions and Critical Phenomena* (Oxford University Press, 2011), p. 20.

something new: there is a dividing point that breaks the line of positive K values into the two above segments. This separator will later be shown to be a particular case of a *critical surface*. The exact solution, contained in the celebrated paper of Onsager[4] to three significant figures, is $K_c = 0.441$.

The above sets the stage for a much extended discussion in later sections.

8.3 RESCALING OF THE TRIANGULAR LATTICE

We now provide a live-teaching example on how to handle (at least in some degree of approximation) the rescaling procedure for a triangular lattice, as depicted in Fig. 8.3A, where the rescaling process can be carried out in closed form. Throughout, we confine ourselves to nearest-neighbor interactions in the ground state.

The RGT (renormalization-group-technique) process involves the relation

$$e^{\hat{\mathcal{H}}'\{S'_\alpha\}} = \sum_{\{S_i\}} e^{\hat{\mathcal{H}}\{S'_\alpha, S_i\}}, \tag{8.31}$$

where the $S_i = \pm 1$ refer to the lattice spins on the vertices of the triangular array of Fig. 8.3A, and the $S'_\alpha = \pm 1, \pm 3$, to the block spin values for each triangle, computed as the sum of the three S_i spins.

We first split the Hamiltonian into a primary portion

$$\hat{\mathcal{H}}_0 = K \sum_\alpha \sum_{i,j\varepsilon\alpha} S_i S_j, \tag{8.32}$$

that deals with spin interactions within a triangle α, and into a perturbing potential

$$\mathcal{V} = K \sum_{\beta \neq \alpha} \sum_{i\in\alpha, j\in\beta} S_i S_j \tag{8.33}$$

for interactions between nearest-neighbor spins in different adjacent blocks. To handle the latter, we introduce the average of $e^{\mathcal{V}}$ weighted by $\hat{\mathcal{H}}_0$ as

$$\left\langle e^{\mathcal{V}} \right\rangle_0 = \frac{\sum_{\{S_i\}} e^{\mathcal{V}\{S'_\alpha, S_i\}} e^{\hat{\mathcal{H}}_0\{S_i\}}}{\sum_{\{S_i\}} e^{\hat{\mathcal{H}}_0\{S_i\}}}. \tag{8.34}$$

We can then rewrite Eq. (8.31) in the form

$$e^{\hat{\mathcal{H}}'\{S'_\alpha\}} = \sum_{\{S_i\}} e^{\hat{\mathcal{H}}_0\{S_i\}} e^{\mathcal{V}\{S'_\alpha, S_i\}} = \left\langle e^{\mathcal{V}} \right\rangle_0 \sum_{\{S_i\}} e^{\hat{\mathcal{H}}_0\{S_i\}}. \tag{8.35}$$

4. L. Onsager, *Phys. Rev.* **65**, 117 (1944).

A) FRAGMENT OF THE TRIANGULAR SPIN LATTICE

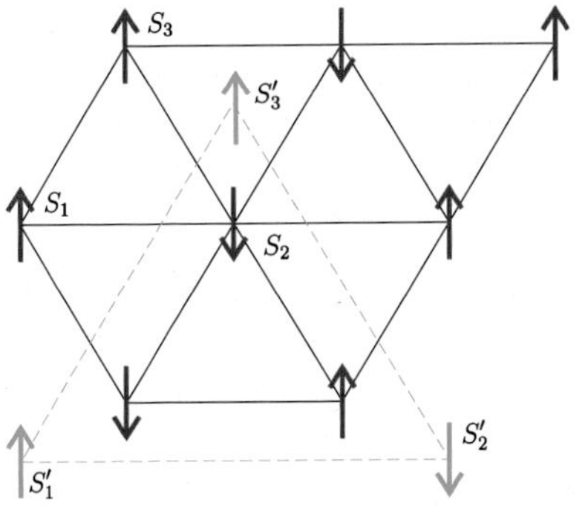

B) SPIN CONFIGURATIONS FOR THE TRIANGULAR SITE

				Interactions			Energy Sum
	1	2	3	12	13	23	
$S'_\alpha = \pm 1$ consistent with	↑	↑	↓	↑↑	↑↓	↑↓	$-K$
	↑	↓	↑	↑↓	↑↑	↓↑	$-K$
	↓	↑	↑	↓↑	↓↑	↑↑	$-K$
$S'_\alpha = +3$	↑	↑	↑	↑↑	↑↑	↑↑	$+3K$
$S'_\alpha = -1$	↓	↓	↑	↓↓	↓↑	↓↑	$-K$
	↓	↑	↓	↓↑	↓↓	↑↓	$-K$
	↑	↓	↓	↑↓	↑↓	↓↓	$-K$
$S'_\alpha = -3$	↓	↓	↓	↓↓	↓↓	↓↓	$+3K$

FIGURE 8.3 A) Portion of a triangular lattice array. The individual spins S on the lattice sites are identified as shown; the block spins are indicated by S'; B) listing of possible spin configurations on a triangular lattice. The concomitant energetic configurations are shown on the right.

The sum on the right extends over N identical triangular units, so that

$$\sum_{\{S_i\}} e^{\hat{\mathcal{H}}_0\{S_i\}} = [\mathcal{Z}_0(K)]^N, \tag{8.36}$$

where $\mathcal{Z}_0$ is the partition function for one block, weighted by $\hat{\mathcal{H}}_0$. Thus,

$$e^{\hat{\mathcal{H}}'\{S'_\alpha\}} = \left\langle e^{\mathcal{V}} \right\rangle_0 [\mathcal{Z}_0(K)]^N. \tag{8.37}$$

We now introduce the *cumulant expansion*

$$\left\langle e^{\mathcal{V}} \right\rangle_0 = 1 + \langle \mathcal{V} \rangle_0 + \frac{1}{2} \left\langle \mathcal{V}^2 \right\rangle_0 + \cdots. \tag{8.38}$$

Now take logarithms and expand the argument in an ascending power series as

$$\ln\left\langle e^{\mathcal{V}}\right\rangle_0 = \langle\mathcal{V}\rangle_0 + \frac{1}{2}\left[\left\langle\mathcal{V}^2\right\rangle_0 - \langle\mathcal{V}\rangle_0^2\right] + \cdots, \tag{8.39}$$

so that taking antilogs, we find that, in this approximation,

$$\left\langle e^{\mathcal{V}}\right\rangle_0 = \exp\left\{\langle\mathcal{V}\rangle_0 + \frac{1}{2}\left[\left\langle\mathcal{V}^2\right\rangle_0 - \langle\mathcal{V}\rangle_0^2\right]\right\}. \tag{8.40}$$

This finally leads us to the relation for the rescaled effective Hamiltonian

$$\hat{\mathcal{H}}'\left\{S'_\alpha\right\} = N\ln\mathcal{Z}_0(K) + \langle\mathcal{V}\rangle_0 + \frac{1}{2}\left[\left\langle\mathcal{V}^2\right\rangle_0 - \langle\mathcal{V}\rangle_0^2\right]. \tag{8.41}$$

We restrict ourselves to the first-order correction; the last term is provided for those brave enough to include second order terms in the ensuing derivation. The term $\langle\mathcal{V}\rangle_0$ relates to the interactions of nearest-neighbor spins on adjacent triangles, such as S_3^β with S_1^α and S_2^α. Accordingly, we set $\mathcal{V} = \sum_{\alpha\neq\beta}\mathcal{V}_{\alpha\beta}$ with

$$\mathcal{V}_{\alpha\beta} = K S_3^\beta(S_1^\alpha + S_2^\alpha). \tag{8.42}$$

Since the two terms in the parentheses are identical, we may write

$$\langle\mathcal{V}_{\alpha\beta}\rangle_0 = 2K\left\langle S_3^\beta S_1^\alpha\right\rangle_0 = 2K\left\langle S_3^\beta\right\rangle_0\left\langle S_1^\alpha\right\rangle_0, \tag{8.43}$$

where the indicated factorization is permissible since the weighting factor $\hat{\mathcal{H}}_0$ does not refer to different blocks.

8.4 DETERMINATION OF AVERAGED PERTURBATION POTENTIALS

We specify the averaged value of a typical lattice site spin via the relation

$$\left\langle S_3^\beta\right\rangle_0 = \frac{1}{\mathcal{Z}_0}\sum_{\left\{S_3^\beta\right\}} S_3^\beta \exp\left\{K\left(S_1^\beta S_2^\beta + S_1^\beta S_3^\beta + S_2^\beta S_3^\beta\right)\right\}, \tag{8.44}$$

where the partition function is read off from the spin summations in the center of Fig. 8.3B, which yield the K values on the right:

$$\mathcal{Z}_0 = 3e^{-K} + e^{3K}. \tag{8.45}$$

For the case listed under $S'_\alpha = 1$, Fig. 8.3B shows that

$$\left\langle S_3^\beta\right\rangle_0 = +\frac{e^{3K} + e^{-K}}{3e^{-K} + e^{3K}}, \tag{8.46}$$

whereas for $S'_\beta = -1$, we encounter

$$\left\langle S_3^\beta \right\rangle_0 = -\frac{e^{3K} + e^{-K}}{3e^{-K} + e^{3K}}. \tag{8.47}$$

In the ground state, the two block spins should be in opposite alignment; hence, we may replace the above plus and minus signs as follows:

$$\left\langle S_3^\beta \right\rangle_0 = S'_\beta \frac{e^{3K} + e^{-K}}{3e^{-K} + e^{3K}} \equiv S'_\beta R(K)$$

and (8.48)

$$\left\langle S^\alpha \right\rangle_0 = S'_\alpha \frac{e^{3K} + e^{-K}}{3e^{-K} + e^{3K}} \equiv S'_\alpha R(K).$$

Invoking Eq. (8.33), we then write

$$\langle \mathcal{V} \rangle_0 = 2K R^2(K) \sum_{\alpha \neq \beta} S'_\alpha S'_\beta, \tag{8.49}$$

which finally allows us to express the transformed Hamiltonian, using Eq. (8.41), in the form

$$\hat{\mathcal{H}}' \left\{ S'_\alpha \right\} = N \ln \mathcal{Z}_0(K) + K' \sum_{\alpha \neq \beta} S'_\alpha S'_\beta \tag{8.50}$$

with

$$K' = 2K R^2(K). \tag{8.51}$$

This achieves what we set out to do! Within our crude approximation scheme, we have determined the scaled Hamiltonian in the format (8.50), which involves the block spins S' in the same functional format as the lattice spins S relate to the lattice Hamiltonian. Involved as well is the partition function for the lattice that does not figure in the RGT operations. Additionally, we encounter the block coupling constant K'; the latter is related to the original coupling constant K as shown in Eq. (8.51). All this conforms to the theoretical outline of Chapter 7.

8.5 DETERMINATION OF THE FIXED POINT

This sets the stage for determining the critical point, at which

$$K^* = 2K^* R^2(K^*), \tag{8.52}$$

with the solutions $K^* = 0, \infty$ or $R(K^*) = {}^1/_{\sqrt{2}}$. The nontrivial K point is most easily found by multiplying $R(K)$ with ${}^{e^K}/_{e^K}$ and solving for K^* numerically;

the result yields

$$K^* = 0.336. \tag{8.53}$$

This may be compared with the exact value of $K = 0.275$ as cited by Onsager for a two-dimensional triangular lattice. Considering the relatively modest theoretical effort involved in the above methodology, the agreement with the exact value is really quite good.

Chapter 9

The Renormalization Group Operations

We come now to one of the key issues in dealing with critical phenomena: studying the so-called real space renormalization technique. *Among other accomplishments, this allows us to place the homogeneity arguments of Chapter 7 on a sounder theoretical basis. Starting again with a lattice of spins, we continue to be guided by the Kadanoff methodology of gradually eliminating microscopic variables through the coarse graining (or blocking) procedure, followed at each step by the reduction of all dimensions by the factor* $b > 1$*, so as to restore the original lattice configuration.*

9.1 REAL SPACE RENORMALIZATION

At the outset, we need to consider some generalities. In executing the blocking methodology, two trivial cases are encountered: if we maintain the temperature of a system above its critical value T_c, then successive rescalings, as already explained, shrink lattice distances and correlation lengths ξ by a factor $b > 1$, so that $\xi' = \frac{\xi}{b}$; after many such operations, this eventually tends to a zero correlation length. This represents the high-T state. Conversely, if we start a system subject to some degree of magnetic order below T_c, then lowering the temperature increases the degree of order at the start of the blocking process. Concurrently, the correlation variables involve ever larger distances until they cover the entire system, indicative of complete order that is unchanged during rescaling. We have reached the low-T attractive critical point.

Clearly, the intermediate case $T = T_c$ is of particular interest; it divides the Hamiltonians for which the various coupling constants K_l (for $T \neq T_c$) of Chapter 8 move the system toward the upper trivial point, from Hamiltonians for which the coupling constants lead the system toward the lower attractive point. This dichotomy is reflected in the existence of a dividing (hyper)plane, called a *critical surface*, spanned by coupling constants K_l as described later. A point on that surface represents a Hamiltonian at a critical temperature, with the other parameters at values corresponding to the location of that point.

A Primer to the Theory of Critical Phenomena. http://dx.doi.org/10.1016/B978-0-12-804685-2.00009-7

9.2 RENORMALIZATION OF THE HAMILTONIAN

We first consider what happens to Hamiltonians during coarse graining. The probability of encountering a particular set of spin assignments $\{S_i\}$ in an Ising lattice is proportional to $\exp\left[-\hat{\mathcal{H}}(\{S_i\})\right]$. Here $\hat{\mathcal{H}}$ is a temperature-dependent *effective Hamiltonian*, either considered as a function or as an operator, as the situation demands. At the outset, we may set $\hat{\mathcal{H}} \equiv \frac{\mathcal{H}}{k_B T}$ with $\mathcal{H}$ independent of T, but after many iterations, the temperature dependence of $\hat{\mathcal{H}}$ may become much more complicated. *To simplify matters, we shall set* $\beta \equiv \frac{1}{k_B T} = 1$.

In line with our earlier discussion, our objective is to render the spin-dependent expectation value $\langle y \rangle$ of an observable the same, whether determined at the nth iteration

$$\begin{aligned}\langle y \rangle &= \frac{1}{\mathcal{Z}_n} \sum_{\left\{S_i^{(n)}\right\}} y\left(\left\{S_i^{(n)}\right\}\right) \exp\left[-\hat{\mathcal{H}}_n\left(\left\{S_i^{(n)}\right\}\right)\right]; \\ \mathcal{Z}_n &= \sum_{\left\{S_i^{(n)}\right\}} \exp\left[-\hat{\mathcal{H}}_n\left(\left\{S_i^{(n)}\right\}\right)\right],\end{aligned} \tag{9.1}$$

or, at the $(n+1)$th iteration,

$$\langle y \rangle = \frac{1}{\mathcal{Z}_{n+1}} \sum_{\left\{S_i^{(n+1)}\right\}} y\left(\left\{S_i^{(n+1)}\right\}\right) \exp\left[-\hat{\mathcal{H}}_{n+1}\left(\left\{S_i^{(n+1)}\right\}\right)\right], \tag{9.2}$$

whereas at the same time requiring that $\hat{\mathcal{H}}_{n+1}$ should have the same functional form as $\hat{\mathcal{H}}_n$.

In general, such a set of conditions cannot be met; some approximations must be invoked to reach this goal as closely as we can. To proceed from $\hat{\mathcal{H}}_n$ to $\hat{\mathcal{H}}_{n+1}$ requires a mapping operation, $H_{n+1} = \hat{\mathcal{M}}_b \hat{\mathcal{H}}_n$, with the operator $\hat{\mathcal{M}}_b$ based on the scaling b, during which the functional form of the Hamiltonian remains invariant but the parameters are altered, a process that was illustrated in Chapter 8.

9.3 TRACKING PARAMETRIC CHANGES DURING COARSE GRAINING

Rather than working with the entire Hamiltonian, we deal with the alterations in the parameters that constitute the Hamiltonian. Hamiltonians of interest generally contain a set of coupling constants $K_1, K_2, \ldots, K_l, \ldots, K_N$ that link the response of a system to various external influences, such as the magnetization of a system by an external field, the reduction in volume when a system is subjected to external pressure, the interaction of spins on adjacent sites in a lattice, etc. These various K_l parameters can be thought of as lying on a set of orthogonal axes along the unit vectors $\hat{\boldsymbol{e}}_l$ that span a *parameter space* $\boldsymbol{K}$. A specific value $K_l^{(j)}$ along each of the l axes ($l = 1, 2, \ldots, N$) then generates a particular point $\boldsymbol{\kappa}$ in space $\boldsymbol{K}$, called a *representative point* for that state of the system. These K_l^j values constitute the components $\left\{K_l^{(j)}\right\}$ of the vector $\boldsymbol{\kappa}$. During the Kadanoff coarse-graining scheme, while keeping functional forms invariant, we execute a set of rescaling operations that change the $K_l^{(j)}$; the representative point changes position and thereby traces out a trajectory. As these components change (or, in common parlance, *flow*) to a different set of coordinates, they become components $\left\{K_l'^{(j')}\right\}$ of a different vector $\boldsymbol{\kappa}'$. This transformation is symbolically represented by an operator $\hat{\mathcal{M}}^{(b)}$ such that

$$\boldsymbol{\kappa}' = \hat{\mathcal{M}}^{(b)} \boldsymbol{\kappa}. \tag{9.3}$$

This process is to be repeated many times because, at each coarse graining stage, we only eliminate a few degrees of freedom. We ultimately seek conditions under which these repetitions no longer produce any change in $\boldsymbol{\kappa}$, corresponding to a *fixed point*, designated as $\boldsymbol{\kappa}^*$.

9.4 AN EXAMPLE

A specific example is the Ising remapping, based on Eqs. (7.1) and (7.4); K and $\mathcal{H}_m$ are the relevant parameters. The corresponding flow patterns are illustrated in Fig. 9.1, which depicts $\boldsymbol{\kappa}$ trajectories in the vicinity of $\boldsymbol{\kappa}^*$. We encounter three possibilities: trajectories that avoid a particular point $\boldsymbol{\kappa}^*$, known as a *repulsive fixed point*; trajectories that converge on a particular point $\boldsymbol{\kappa}^*$, known as an *attractive fixed point*; trajectories, some of which move toward, and others away from, a particular point $\boldsymbol{\kappa}^*$, known as a *mixed point*. Any fixed point is characterized by the relation $\boldsymbol{\kappa}^* = \hat{\mathcal{M}}_b \boldsymbol{\kappa}^*$, so that $\boldsymbol{\kappa}^*$ is mapped onto itself. If under coarse graining, $\boldsymbol{\kappa}$ remains on a critical surface, it is subject to several possibilities: (i) $\boldsymbol{\kappa}$ tends to a finite limit $\boldsymbol{\kappa}^*$; (ii) $\boldsymbol{\kappa}$ tends to infinity; (iii) $\boldsymbol{\kappa}$ moves around on the surface in cyclic pattern without tending to a finite limit $\boldsymbol{\kappa}^*$; (iv) $\boldsymbol{\kappa}$ erratically wanders on the surface without ever reaching a destination $\boldsymbol{\kappa}^*$. Here only case (i) is of interest. Our important future project is the identification of *fixed points* $\boldsymbol{\kappa}^*$ on the *critical surface* that are mixed: we can then focus on a limited number of concomitant repulsive trajectories that move the system away from the critical surface.

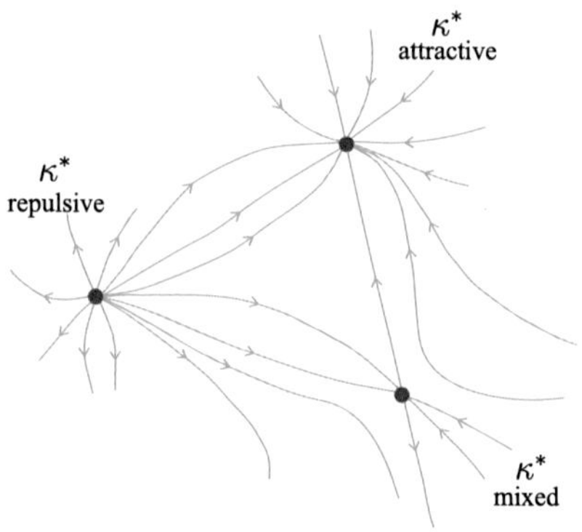

FIGURE 9.1 Schematic diagram showing various two-dimensional trajectories converging on or avoiding three fixed points. These are labeled as shown.

A mathematical analysis of these trajectories is based on a study of the immediate vicinity $\delta\boldsymbol{\kappa}$ of $\boldsymbol{\kappa}^*$, where a linear approximation suffices. That is, using $\hat{\boldsymbol{\mathcal{M}}}^{(b)}$ to operate on the function $\boldsymbol{\kappa}^* + \delta\boldsymbol{\kappa}$, to yield $\boldsymbol{\kappa}^* + \delta\boldsymbol{\kappa}'$, we write

$$\hat{\boldsymbol{\mathcal{M}}}^{(b)}(\boldsymbol{\kappa}^* + \delta\boldsymbol{\kappa}) = \boldsymbol{\kappa}^* + \delta\boldsymbol{\kappa}' = \boldsymbol{\kappa}' = \hat{\boldsymbol{\mathcal{M}}}^{(b)}\boldsymbol{\kappa}^* + \hat{\boldsymbol{\mathcal{M}}}^{(b)}\delta\boldsymbol{\kappa} = \boldsymbol{\kappa}^* + \hat{\boldsymbol{\mathcal{M}}}^{(b)}\delta\boldsymbol{\kappa}. \tag{9.4}$$

Componentwise, this involves the linear term in a set of Taylor expansions given by

$$K_i' = K_i^* + \sum_n \left[\frac{\partial K_i'}{\partial K_n}\right]_{K_n = K_n^*} \delta K_n + \cdots \quad (n = 1, 2, \ldots, N). \tag{9.5}$$

Thus, for given i and n,

$$\delta K_i' = \mathcal{M}_{in}^{(b)} \delta K_n; \quad \mathcal{M}_{in}^{(b)} \equiv \left[\frac{\partial K_i'}{\partial K_n}\right]_{K_n = K_n^*}^{(b)}, \tag{9.6}$$

where b is the usual scaling factor. These $\mathcal{M}_{in}^{(b)}$ constitute the entries of the matrix $\hat{\boldsymbol{\mathcal{M}}}^{(b)}$.

For further progress, assume that $\hat{\boldsymbol{\mathcal{M}}}^{(b)}$ is a symmetric matrix that can be diagonalized, as symbolized by $\hat{\boldsymbol{\mathcal{M}}}_d^{(b)}$; standard procedures, illustrated by a simple example in the Appendix A, result in equations featuring a set of corresponding eigenvalues $\lambda_l^{(b)}$ and eigenfunctions $\hat{\boldsymbol{e}}_l$. Here l labels the N distinct eigenvalues derived from the diagonalization process. The number of basis vectors matches the required number of K_l parameters in the Hamiltonian.

On completion of the diagonalization, we end up with relations in the guise of N *eigenvalue equations* of the general form

$$\hat{\mathcal{M}}_d^{(b)}(\kappa^*)\hat{e}_l = \lambda_l^{(b)}\hat{\boldsymbol{e}}_l \tag{9.7}$$

with the tacit assumption that the $\lambda_l^{(b)}$ are nondegenerate. For a live demonstration of the process, consult the Appendix A.

Let us note what has been accomplished so far. The above rescaling operations not only alter the location of $\boldsymbol{\kappa}$ but also successively reduce the number of variables needed to describe these properties, as prescribed by Eqs. (9.1) and (9.2). The resulting changes are analogous to symmetry operations, chosen so that the system remains (perhaps, only approximately) invariant. Correspondingly, the process is designed so that the average involving the transformed variables over the reformulated probabilities is not changed relative to the average of the nontransformed variables over the original probabilities. This ensures that the physical characteristics of the lattice remain unchanged.

9.5 GROUP PROPERTIES OF THE COARSE GRAINING OPERATION

We next attend to additional requirements. As is physically reasonable, in executing iterative procedures, we demand that two successive scaling operations, b_1 followed by b_2, should produce the same effect as a single one step operation b_2b_1: correspondingly, we write $\hat{\mathcal{M}}^{(b2)}\hat{\mathcal{M}}^{(b1)} = \hat{\mathcal{M}}^{(b21)}$. This is one of the requirements for elements that form a *semigroup* in mathematical operations.[1] In addition, the $\hat{\mathcal{M}}^{(b)}$ operations must reflect the symmetry of the lattice and must fulfill the identity requirement, and hence is called the *renormalization group transformation* (RGT). Of particular interest are trajectories that converge to a *fixed point* (or to one of several such fixed points), depending on the particular system that is being studied, and where one starts off in space $\boldsymbol{K}$. A schematic two-dimensional representation of such dynamics in that space is provided in Fig. 9.1. See more on this below.

The same semigroup property of $\hat{\mathcal{M}}_d^{(b)}$ must also be obeyed by the eigenvalues of Eq. (9.7): $\lambda_l(b2)\lambda_l(b1) = \lambda_l(b21)$. Although this is almost self-evident,

1. We deal here with a semigroup rather than with a full group because the inverse to any operation is absent; during the coarse graining processes, some details of the system state are lost.

we provide a proof by explicitly considering the sequence

$$\hat{\mathcal{M}}_d^{(b21)} \hat{\boldsymbol{e}}_l = \lambda_l(b21)\hat{\boldsymbol{e}}_l = \hat{\mathcal{M}}_d^{(b2)} \left[\hat{\mathcal{M}}_d^{(b1)} \hat{\boldsymbol{e}}_l\right] = \lambda_l(b1)\left[\hat{\mathcal{M}}_d^{(b2)} \hat{\boldsymbol{e}}_l\right] = \lambda_l(b2)\lambda_l(b1)\hat{\boldsymbol{e}}_l, \tag{9.8}$$

which verifies the claim.

9.6 SCALING FIELDS AND PROPERTIES

What is the importance of the above analysis? The condition $\lambda_l(b2)\lambda_l(b1) = \lambda_l(b21)$ leads directly to the identification

$$\lambda_l(b) = b^{y_l}, \tag{9.9}$$

which is the sought-after, all important, power law dependence of λ on $b > 1$ and involves the *critical index* y_l as a quantity governing critical behavior. As we will see, this ultimately leads us to the homogeneity relations.

With $\hat{\boldsymbol{e}}_l$ serving as the relevant basis vectors that span the field of the K_l (Hamiltonian) parameters, we expand $\delta\boldsymbol{\kappa}$ as

$$\delta\boldsymbol{\kappa} = \sum_l g_l \hat{\boldsymbol{e}}_l. \tag{9.10}$$

Consider next the following expansion of the scaling operation $\delta\boldsymbol{\kappa}' = \hat{\mathcal{M}}_d^{(b)}(\boldsymbol{\kappa}^*)\delta\boldsymbol{\kappa}$:

$$\delta\boldsymbol{\kappa}' = \sum_l g_l' \hat{\boldsymbol{e}}_l = \sum_l \hat{\mathcal{M}}_d^{(b)} g_l \hat{\boldsymbol{e}}_l = \sum_l g_l \lambda_l \hat{\boldsymbol{e}}_l = \sum_l b^{y_l} g_l \hat{\boldsymbol{e}}_l, \tag{9.11}$$

where the $b^{y_l} g_l$ are termed *scaling fields*, for which

$$g_l' = b^{y_l} g_l. \tag{9.12}$$

In summary, scaling variables characterize the parameter space near the critical point. As an example, for ferromagnets, we may identify two g_l parameters, $t \equiv \frac{T-T_c}{T_c}$ and $h \equiv \frac{H_m - H_{mc}}{H_{mc}}$. We learn about critical phenomena through the study of the fixed points, the eigenvalues or exponents y_l in Eq. (9.9), and the scaling fields $b^{y_l} g_l$ in Eq. (9.12). We now examine these changes in more detail.

9.7 CLASSIFICATION OF VARIABLES AND RELATED FIXED POINTS

The relation $\delta\boldsymbol{\kappa}' = \hat{\mathcal{M}}_d^{(b)}(\boldsymbol{\kappa}^*)\delta\boldsymbol{\kappa}$, rewritten in the form (9.11), merits careful scrutiny: The various g_l' represent the components of the vector $\delta\boldsymbol{\kappa}'$ along the various orthogonal directions $\hat{\boldsymbol{e}}_l$. Each involves the product $g_l\lambda_l$. We can then arrange the eigenvalues in increasing order of their absolute values: $|\lambda_1| < |\lambda_2| < |\lambda_3| < \cdots$. Some of these grow, and others shrink, as the $\hat{\mathcal{M}}_d^{(b)}(\boldsymbol{\kappa}^*)$ operation is carried out. We are then faced with three possibilities:

(i) $|\lambda_l| > 1$, so that y_l is positive ($y_l > 0$) and $b^{y_l} > 1$; then g_l' increases with each iteration, and the system moves away from the critical point in the direction of the eigenvector $\hat{\boldsymbol{e}}_l$. The corresponding scaling field variable g_l' is said to be a *relevant variable*. The matching fixed point was earlier identified as *repulsive* (cf. Fig. 9.1). If we are looking for critical points, then we must learn how to attend to this problem.

(ii) $|\lambda_l| < 1$, so that y_l is negative ($y_l < 0$) and $b^{y_l} < 1$; then g_l' decreases with each iteration, and the system under such operations automatically approaches the critical point in the direction of the eigenvector $\hat{\boldsymbol{e}}_l$. Such g_l' is said to be an *irrelevant variable*. The corresponding fixed point was earlier said to be *attractive*.

(iii) $|\lambda_l| = 1$, so that $y_l = 0$. The corresponding g_l' is said to be *marginal*, in the sense that g_l' remains unchanged. This case must be treated by going to higher-order terms in the expansion.

Thus, if we start at a point $\boldsymbol{\kappa}$ close to $\boldsymbol{\kappa}^*$ but not on the critical manifold, then relevant variables drive the system further away from $\boldsymbol{\kappa}^*$, whereas irrelevant variables lead the system toward the fixed point. Clearly, the former set of variables must be attended to if we are to reach the fixed point by closing in on the critical surface. Note that the above designations (attractive, repulsive, mixed) must always be taken with respect to a specific fixed point. The three types of fixed points are illustrated in Fig. 9.1.

The above designations are perhaps best understood by considering a particular case of Eq. (9.11):

$$\delta\boldsymbol{\kappa}' = b^{y_1} g_1 \hat{\boldsymbol{e}}_1 + b^{y_2} g_2 \hat{\boldsymbol{e}}_2, \tag{9.13}$$

where we associate the first term with temperature changes and the second with properties such as magnetic fields that diminish under rescaling. For T very close to T_c, we can reasonably include temperature variations by setting $g_1 = |g_1^0 \cdot (T - T_c)|$, whence

$$\delta\boldsymbol{\kappa}' = b^{y_1} |g_1^0 \cdot (T - T_c)| \hat{\boldsymbol{e}}_1 + b^{y_2} g_2 \hat{\boldsymbol{e}}_2. \tag{9.14}$$

Now introduce the correlation lengths that diverge as the critical point is approached, as specified by the standard expression

$$\xi = |g_1^0 \cdot (T - T_c)|^{-1/y_1}, \tag{9.15}$$

whereby

$$\delta\boldsymbol{\kappa}' = \left(\frac{b}{\xi}\right)^{y_1} \hat{\boldsymbol{e}}_1 + b^{y_2} g_2 \hat{\boldsymbol{e}}_2. \tag{9.16}$$

By assumption $y_2 < 0$, so that the second term automatically approaches zero on repeated rescaling, which is the effect of an irrelevant exponent. However, in the first term, we have to set $y_1 > 0$ because correlation lengths necessarily increase as $T \to T_c$; we deal with a relevant exponent. To actually reach the critical point, $\delta\boldsymbol{\kappa}' \to 0$, we must therefore let T coincide with T_c. The fixed point then involves one relevant and one irrelevant critical index and thus of the mixed type. Incidentally, by convention we set $\nu \equiv \frac{1}{y_1}$, whereby the divergence of the correlation length is rewritten in the form

$$\xi \sim |(T - T_c)|^{-\nu}. \tag{9.17}$$

9.8 BACK TO HOMOGENEITY RELATIONS

We now find out how the $g_l' = b^{y_l} g_l$ scaling operation is utilized. At an elementary level, consider a small region of space characterized by a free-energy density f. This physical quantity must remain the same, whether described at the level of the nth or $(n+1)$th iteration. The rescaling of the free energy of that volume element is thus subject to the rule

$$f(\hat{\mathcal{H}}')d^d x' = f(\hat{\mathcal{H}})d^d x = f(\hat{\mathcal{H}}')b^{-d} d^d x, \tag{9.18}$$

where on the right we scaled the $d^d x'$ position variable. It follows that

$$f(\hat{\mathcal{H}}) = b^{-d} f(\hat{\mathcal{H}}'). \tag{9.19}$$

The rule (9.19) applies to all the coupling constants that occur as parameters in the Hamiltonian and that form the components g_l of the $\boldsymbol{\kappa}$ vector: hence,

$$f(\boldsymbol{\kappa}) = b^{-d} f(\boldsymbol{\kappa}'). \tag{9.20}$$

Thus, if the free energy density involves the scaling variables $b^{y_l} g_l$ as parameters, Eqs. (9.10) and (9.12), then, close to the fixed point, relation (9.20) reads

$$f(g_1, g_2, \ldots) = b^{-d} f(b^{y_1} g_1, b^{y_2} g_2, \ldots). \tag{9.21}$$

We have recovered the generalized homogeneity requirement that was introduced in Chapter 7 on an intuitive basis. This particular relation is of special importance because all thermodynamic quantities of interest can be derived from the Helmholtz free energy or from other corresponding function of state.

We now proceed as follows: Consider the general problem of studying the free energy changes contingent on the renormalization. In general the Boltzmann factors for the Hamiltonians at stages n and $n+1$ in the renormalization are related in the following generic manner:

$$\exp\left[-\hat{\mathcal{H}}^{(n+1)}\left(\left\{S_\alpha^{(n+1)}\right\}\right)\right]=\exp[-G]\cdot\sum_{\left\{S_i^{(n)}\right\}}\exp\left[-\hat{\mathcal{H}}^{(n)}\left(\left\{S_i^{(n)}\right\}\right)\right], \quad (9.22)$$

where the summation extends over all sets $\left\{S_i^{(n)}\right\}$ consistent with the sets $\left\{S_\alpha^{(n+1)}\right\}$. The function G need not be explicitly specified, other than requiring that it be independent of the set $S_\alpha^{(n+1)}$ and that it satisfy whatever other requirements the Hamiltonians are supposed to meet.

Now sum Eq. (9.22) over all $S_\alpha^{(n+1)}$ configurations to generate the partition functions; then take logarithms to find the corresponding free energies ($\beta=1$):

$$F^{(n+1)}\left[\hat{\mathcal{H}}^{(n+1)}\left(\left\{S_\alpha^{(n+1)}\right\}\right)\right]=F^{(n)}\left[\hat{\mathcal{H}}^{(n)}\left(\left\{S_i^{(n)}\right\}\right)\right]-G\left[\hat{\mathcal{H}}^{(n)}\left(\left\{S_i^{(n)}\right\}\right)\right]. \quad (9.23)$$

As stated earlier, rather than dealing with Hamiltonians, we can introduce the set of parameters K_i that are collectively represented by the vector $\boldsymbol{\kappa}$. Then introduce the free energy densities

$$\begin{aligned} f^{(n)}\left(\boldsymbol{\kappa}^{(n)}\right)&=\frac{F^{(n)}\left(\boldsymbol{\kappa}^{(n)}\right)}{N^{(n)}},\\ f^{(n+1)}\left(\boldsymbol{\kappa}^{(n+1)}\right)&=\frac{F^{(n+1)}\left(\boldsymbol{\kappa}^{(n+1)}\right)}{N^{(n+1)}}=\frac{F^{(n+1)}\left(\boldsymbol{\kappa}^{(n+1)}\right)}{\frac{N^{(n)}}{b^d}},\\ \text{and}\quad g(\boldsymbol{\kappa}^{(n)})&=\frac{G}{N^{(n)}}, \end{aligned} \quad (9.24)$$

so that

$$f^{(n+1)}\left(\boldsymbol{\kappa}^{(n+1)}\right)=b^d\left[f^{(n)}\left(\boldsymbol{\kappa}^{(n)}\right)-g\left(\boldsymbol{\kappa}^{(n)}\right)\right]. \quad (9.25)$$

At this stage, we introduce the so-called *singularity assumption*, that the various divergences encountered close to the critical point arise through the functions f, without involving $g(\boldsymbol{\kappa}^{(n)})$. This may not always be the case; for further discussion of this delicate point, you need to consult the literature. Here we proceed under the assumption that $g(\boldsymbol{\kappa}^{(n)})$ plays no further role. Then by representing $\boldsymbol{\kappa}^{(n)}$ and $\boldsymbol{\kappa}^{(n+1)}$ through the expansion coefficients (9.10) and (9.12) we recover Eq. (9.21).

We can do more: for the special case of a system with two degrees of freedom, we can adjust the arbitrary scaling factor to take the convenient value $b = g_1^{-1/y_1}$. In that event, we find that

$$f(g_1, g_2) = g_1^{\frac{d}{y_1}} f\left(1, \frac{g_2}{g_1^{y_2/y_1}}\right). \tag{9.26}$$

This is a restatement of, and justification for, the *Widom scaling hypothesis* that we dealt with in Chapter 7. Near the critical point, the binary thermodynamic functions can be rewritten so as to involve solely one degree of freedom as an independent variable of a universal function f. Conversely, the Widom scaling operation (9.26) implies the existence of a homogeneity relation (9.21) for two degrees of freedom.

9.9 CONSEQUENCES

Let us consider the above topic at a less formal level. In a magnetic system for which the applicable parameters $\{g_l\}$ are the reduced temperature $t = \frac{T-T_c}{T_c}$ and the scaled magnetic field $h = \frac{H_m - H_{mc}}{H_{mc}}$, we identify $g_1 \equiv t$ and $g_2 \equiv h$, so that close to the critical point $t = h = 0$, both $t \to 0$ and $h \to 0$. In fact, Eq. (9.21) reduces to

$$f(t,h) = b^{-d} f(b^{y_1} t, b^{y_2} h). \tag{9.27}$$

Now examine the approach of the heat capacity at constant h to the critical point, as specified in the standard form

$$C_h = -T_c \left(\frac{\partial^2 f}{\partial t^2}\right)_{h=0} = |t|^{-\alpha}, \tag{9.28}$$

and write out the second derivative of Eq. (9.21) as

$$\left(\frac{\partial^2 f}{\partial t^2}\right)_{h=0} = b^{-d+2y_1} f''(b^{y_1} t, 0). \tag{9.29}$$

Then adopt the Widom scaling method by setting $b = |t|^{-1/y_1}$, whereby

$$C_h \sim |t|^{\frac{d-2y_1}{y_1}} f''(1,0). \tag{9.30}$$

Comparison with Eq. (9.28) shows that

$$\alpha = 2 - \frac{d}{y_1}, \tag{9.31}$$

which allows us to fix the y_1 parameter in terms of a measurable critical index. We had earlier set $\xi \sim |t|^{-\nu}$. All distances rescale according to the adopted pattern $b = |t|^{-\frac{1}{y_1}}$, which allows us to set $\nu = \frac{1}{y_1}$ in general, whence

$$\alpha = 2 - d\nu, \tag{9.32}$$

which is known as *Josephson's law* or as the *hyperscaling relation* (though it is not clear what is *hyper* about it).

APPENDIX A

By way of illustration we consider here the essentials of the matrix diagonalization procedure for a particularly simple case. Consider a problem in linear algebra of the following type:

$$\begin{aligned} h_{11}u_1 + h_{12}u_2 &= \lambda u_1, \\ h_{21}u_1 + h_{22}u_2 &= \lambda u_2, \end{aligned} \tag{9.33}$$

or, in matrix notation,

$$\begin{bmatrix} h_{11} & h_{12} \\ h_{21} & h_{22} \end{bmatrix} \begin{pmatrix} u_1 \\ u_2 \end{pmatrix} = \lambda \begin{pmatrix} u_1 \\ u_2 \end{pmatrix} \quad \text{or} \quad \begin{bmatrix} h_{11} - \lambda & h_{12} \\ h_{21} & h_{22} - \lambda \end{bmatrix} \begin{pmatrix} u_1 \\ u_2 \end{pmatrix} = \begin{pmatrix} 0 \\ 0 \end{pmatrix}, \tag{9.34}$$

which is symbolically rewritten as

$$\hat{\boldsymbol{H}} u = \lambda u, \quad \hat{\boldsymbol{H}} \equiv \begin{bmatrix} h_{11} & h_{12} \\ h_{21} & h_{22} \end{bmatrix}. \tag{9.35}$$

Here u is a column vector with vector components $u_1 = u_1|\alpha\rangle$ and $u_2 = u_2|\beta\rangle$, where $|\alpha\rangle$ and $|\beta\rangle$ are orthogonal unit vectors (kets) satisfying the relations $\langle\alpha|\alpha\rangle = \langle\beta|\beta\rangle = 1$ and $\langle\alpha|\beta\rangle = \langle\beta|\alpha\rangle = 0$. The problem is to determine λ and u from the known h_{ij}.

The procedure depends on the fact that a nontrivial solution of Eq. (9.34) exists if and only if the related determinant vanishes, that is,

$$\begin{vmatrix} h_{11} - \lambda & h_{12} \\ h_{21} & h_{22} - \lambda \end{vmatrix} = 0. \tag{9.36}$$

For illustrative purposes, assume that $h_{11} = h_{22} = a$ and $h_{12} = h_{21} = b$. We can now set up the following relations:

$$\begin{aligned} \hat{\boldsymbol{H}}|\alpha\rangle &= a|\alpha\rangle + b|\beta\rangle, \\ \hat{\boldsymbol{H}}|\beta\rangle &= a|\beta\rangle + b|\alpha\rangle, \end{aligned} \tag{9.37}$$

from which we form four matrix elements:

$$\langle\alpha|\hat{\boldsymbol{H}}|\alpha\rangle = \langle\beta|\hat{\boldsymbol{H}}|\beta\rangle = a, \quad \langle\beta|\hat{\boldsymbol{H}}|\alpha\rangle = \langle\beta|\hat{\boldsymbol{H}}|\alpha\rangle = b, \tag{9.38}$$

so that the operator in matrix form reads

$$\hat{\boldsymbol{H}} = \begin{bmatrix}\langle\alpha|\langle\beta|\end{bmatrix}\begin{bmatrix} a & b \\ b & a \end{bmatrix}\begin{bmatrix}|\alpha\rangle \\ |\beta\rangle\end{bmatrix}. \tag{9.39}$$

The corresponding determinantal equation is specified by

$$\begin{vmatrix} a-\lambda & b \\ b & a-\lambda \end{vmatrix} = 0 \tag{9.40}$$

with the expansion

$$(a-\lambda)^2 - b^2 = 0, \tag{9.41}$$

yielding the solutions

$$\lambda_+ = a+b, \quad \lambda_- = a-b. \tag{9.42}$$

We now look for the corresponding eigenfunctions, which involve construction of the following linear combinations:

$$\begin{aligned} |+\rangle &= a_+|\alpha\rangle + b_+|\beta\rangle \text{ associated with } \lambda_+, \\ |-\rangle &= a_-|\alpha\rangle + b_-|\beta\rangle \text{ associated with } \lambda_-. \end{aligned} \tag{9.43}$$

To determine the above coefficients, start with Eq. (9.34):

$$\begin{aligned} (h_{11} - \lambda_+)a_+ + h_{12}b_+ &= 0, \\ h_{21}a_+ + (h_{22} - \lambda_+)b_+ &= 0. \end{aligned} \tag{9.44}$$

Then substituting Eq. (9.42) into system (9.44), we find that both equations of the latter lead to the same result,

$$-ba_+ + bb_+ = 0, \quad \text{or} \quad a_+ = b_+. \tag{9.45}$$

Next, impose the normalization requirement

$$\langle +|+\rangle = 1 = a_+^2\langle\alpha|\alpha\rangle + a_+b_+\langle\alpha|\beta\rangle + b_+a_+\langle\beta|\alpha\rangle + b_+^2\langle\beta|\beta\rangle = a_+^2 + b_+^2. \tag{9.46}$$

Thus, selecting the positive root, we find that

$$a_+ = b_+ = \frac{1}{\sqrt{2}} \tag{9.47}$$

and

$$|+\rangle = \frac{1}{\sqrt{2}}\left(|\alpha\rangle + |\beta\rangle\right). \tag{9.48}$$

Similar reasoning leads to

$$|-\rangle = \frac{1}{\sqrt{2}}\left(|\alpha\rangle - |\beta\rangle\right). \tag{9.49}$$

We have now found both the eigenvalues and the corresponding eigenfunctions of Eq. (9.37). You may readily check that the above results still hold for any multiple of the eigenvectors.

An important point of note concerns the diagonalization of matrices. Carrying out the matrix multiplications

$$\begin{bmatrix} \frac{1}{\sqrt{2}} & \frac{1}{\sqrt{2}} \\ \frac{1}{\sqrt{2}} & -\frac{1}{\sqrt{2}} \end{bmatrix} \begin{bmatrix} a & b \\ b & a \end{bmatrix} \begin{bmatrix} \frac{1}{\sqrt{2}} & \frac{1}{\sqrt{2}} \\ \frac{1}{\sqrt{2}} & -\frac{1}{\sqrt{2}} \end{bmatrix}$$
$$= \frac{1}{2}\begin{bmatrix} 1 & 1 \\ 1 & -1 \end{bmatrix} \begin{bmatrix} a & b \\ b & a \end{bmatrix} \begin{bmatrix} 1 & 1 \\ 1 & -1 \end{bmatrix} = \begin{bmatrix} a+b & 0 \\ 0 & a-b \end{bmatrix}, \tag{9.50}$$

we convert the original matrix (9.39) to the diagonal form whose entries match the eigenvalues listed in Eq. (9.42). This illustrates the general theorem that a matrix of the present genre can be diagonalized by carrying out a similarity transformation in which the postfactor, and its transpose as a prefactor, are constructed by adopting the eigenvector components as column entries. The diagonal entries list the corresponding eigenvalues.

Examples

We now show, by means of two examples, how the above method is implemented in practice.

We shall later have an occasion to solve equations of the form

$$\begin{bmatrix} b^2 & B(b^2 - b^{\epsilon}) \\ 0 & b^{\epsilon} \end{bmatrix} \begin{pmatrix} r \\ u \end{pmatrix} = \begin{pmatrix} 0 \\ 0 \end{pmatrix}, \tag{9.51}$$

which leads to a nontrivial solution iff the corresponding determinant vanishes, that is, if

$$\begin{vmatrix} b^2 - \lambda & B(b^2 - b^{\epsilon}) \\ 0 & b^{\epsilon} - \lambda \end{vmatrix} = 0. \tag{9.52}$$

An expansion leads to two distinct eigenvalues $\lambda_1 = b^\epsilon$ and $\lambda_2 = b^2$, distinct because the eigenvalues would otherwise be degenerate, requiring a different approach. Equation (9.52) may be rewritten as

$$\begin{aligned}(b^2-\lambda)r + B(b^2-b^\epsilon)u &= 0,\\ (b^\epsilon-\lambda)u &= 0.\end{aligned} \tag{9.53}$$

Since the resulting vector with components r and u is not known to within an arbitrary positive or negative multiplier, we find it convenient to set $r = 1$.

Now adopt $\lambda_1 = b^\epsilon$. Then the second equation is indeterminate, but the first reduces to

$$(b^2 - b^\epsilon)[1 + Bu] = 0. \tag{9.54}$$

Because of nondegeneracy, the first factor cannot be allowed to vanish; hence, $u = -\frac{1}{B.}$ The resulting vector with components $(1, -\frac{1}{B})$ can be specified only within an arbitrary factor; hence, we are free to multiply both components by $-B$ to generate a vector $\hat{\boldsymbol{e}}^{(2)}$ with components $(-B, 1)$, corresponding to the eigenvalue $\lambda_1 = b^\epsilon$.

Now adopt $\lambda_2 = b^2$. The resulting equations read

$$\begin{aligned}(b^2-b^\epsilon)u &= 0,\\ (b^\epsilon-b^2)u &= 0.\end{aligned} \tag{9.55}$$

Since neither B nor $(b^\epsilon - b^2)$ can be allowed to vanish, we are forced to set $u = 0$. We thereby generate the eigenvector $\hat{\boldsymbol{e}}^{(1)}$ with components $(1, 0)$.

The upshot is that a point $\boldsymbol{K} = r\boldsymbol{r}_0 + u\boldsymbol{u}_0$ was originally specified in the space $\boldsymbol{\kappa}$ via the unit vectors $\boldsymbol{r}_0$ and $\boldsymbol{u}_0$. The transformed unit vectors are now given by $\hat{\boldsymbol{e}}^{(1)} = \boldsymbol{r}_0$ and $\hat{\boldsymbol{e}}^{(2)} = -B\boldsymbol{r}_0 + \boldsymbol{u}_0$. Then, in terms of the new set, we find that $\boldsymbol{K} = (r + uB)\hat{\boldsymbol{e}}^{(1)} + u\hat{\boldsymbol{e}}^{(2)}$. The axes defined by $\hat{\boldsymbol{e}}^{(1)}$ and $\hat{\boldsymbol{e}}^{(2)}$ converge directly to the fixed point $\boldsymbol{K}^*$ at the origin.

As a second example, we adopt an illustration by McComb.[2] Going back to Chapter 8, consider again the two-dimensional magnet characterized by spin interactions K and L among nearest-neighbor and next nearest-neighbor pairs, two coupling constants that take the place of K_1 and K_2. We adopt the following interrelations cited for the change in coupling constant in the course of one RGT:

$$K' = 2K^2 + L \quad \text{and} \quad L' = K^2. \tag{9.56}$$

At the fixed point, this involves the relations

$$K^* = 2K^{*2} + L^* \quad \text{and} \quad L^* = K^{*2}. \tag{9.57}$$

2. W.D. McComb, *Renormalization Methods* (Clarendon Press, Oxford, UK, 2004), pp. 194 ff.

It is a simple matter to establish that this gives rise to three different fixed points: $(K^*, L^*) = (0, 0)$, (∞, ∞), and $(\frac{1}{3}, \frac{1}{9})$. We linearize about the latter, nontrivial case, setting

$$K' = K^* + \delta K', \quad L' = L^* + \delta L'; \\ K = K^* + \delta K, \quad L = L^* + \delta L. \tag{9.58}$$

Now differentiate Eqs. (9.56) and retain only the first-order terms:

$$\delta K' = 4K\delta K + \delta L \approx 4K^*\delta K + \delta L; \\ \delta L' = 2K\delta K \approx 2K^*\delta K. \tag{9.59}$$

These two equations may be recast in matrix form as

$$\begin{bmatrix} \delta K' \\ \delta L' \end{bmatrix} = \begin{bmatrix} 4K^* & 1 \\ 2K^* & 0 \end{bmatrix} \begin{bmatrix} \delta K \\ \delta L \end{bmatrix} = \begin{bmatrix} \frac{4}{3} & 1 \\ \frac{2}{3} & 0 \end{bmatrix} \begin{bmatrix} \delta K \\ \delta L \end{bmatrix} \tag{9.60}$$

for the immediate vicinity of the fixed point. We examine the corresponding determinantal relation

$$\begin{vmatrix} \frac{4}{3} - \lambda & 1 \\ \frac{2}{3} & -\lambda \end{vmatrix} = 0, \tag{9.61}$$

from which we obtain

$$\lambda^2 - \frac{4}{3}\lambda - \frac{2}{3} = 0 \tag{9.62}$$

with two roots

$$\lambda_{1,2} = \frac{1}{3}(2 \pm \sqrt{10}), \quad \text{or} \quad \lambda_1 = 1.721, \ \lambda_2 = -0.387. \tag{9.63}$$

Notice that $|\lambda_1| > 1$ and $|\lambda_2| < 1$, so that we deal here with a mixed fixed point of the type discussed earlier. The vectors are found by inserting the two eigenvalues in

$$\begin{bmatrix} \frac{4}{3} - \lambda & 1 \\ \frac{2}{3} & -\lambda \end{bmatrix} \begin{bmatrix} x \\ y \end{bmatrix} = \begin{bmatrix} 0 \\ 0 \end{bmatrix}. \tag{9.64}$$

Executing the matrix multiplications, we obtain

$$\begin{bmatrix} -0.388x + y \\ 0.667x - 1.721y \end{bmatrix} = \begin{bmatrix} 0 \\ 0 \end{bmatrix} \quad \text{and} \quad \begin{bmatrix} 1.721x + y \\ 0.667x + 0.387y \end{bmatrix} = \begin{bmatrix} 0 \\ 0 \end{bmatrix}. \tag{9.65}$$

As before, the components x, y are subject to sign ambiguities and are known only to within a common arbitrary multiplying factor. This allows us to select

$x = 1$, so that the two vectors are given by

$$\hat{\boldsymbol{e}}^{(1)} = \begin{bmatrix} 1 \\ 0.388 \end{bmatrix}, \quad \hat{\boldsymbol{e}}^{(2)} = \begin{bmatrix} 1 \\ -1.721 \end{bmatrix}, \tag{9.66}$$

respectively. The above components x and y specify the directions of approach to the fixed point at $\left(\frac{1}{3}, \frac{1}{9}\right)$.

Chapter 10

Additional Interrelations Between Critical Exponents

At this stage, we develop additional interrelations among the critical exponents; some earlier material is reviewed to render the discussion more self-contained. To facilitate comparison with the literature, we also introduce a new parameter ω_c that we later eliminate.

10.1 MAGNETIZATION AND CORRELATION EFFECTS

In terms of the scaling parameter b, the interaction from stage n to stage $n+1$ when rescaling spins S_i to block spins S'_α is conventionally specified by

$$S_\alpha^{(n+1)} = b^{-\omega_c d} \sum_{i \in \alpha} S_i^{(n)}, \tag{10.1}$$

where for a fixed d, we allow ω_c to depend on the temperature and other parameters of the system. Then the magnetization at stage n is determined via $M^{(n)} \equiv \left\langle S_i^{(n)} \right\rangle$ and is related to that of stage $n+1$ by

$$M^{(n+1)} = \left\langle S_\alpha^{(n+1)} \right\rangle = b^{-\omega_c d} \sum_{i \varepsilon \alpha} \left\langle S_i^{(n)} \right\rangle. \tag{10.2}$$

In a translationally uniform system, each of the b^d terms in the sum is the same, so that the sum is replaced $b^d M^{(n)}$, whence

$$M^{(n+1)} = b^{d(1-\omega_c)} M^{(n)}. \tag{10.3}$$

As introduced in Chapter 3, Eq. (3.19), the correlation functions involve the spin variables as

$$G^{(n)}(i, j) = \left\langle S_i^{(n)} S_j^{(n)} \right\rangle, \tag{10.4}$$

whence, according to Eq. (10.1), for block variables,

$$G^{(n+1)}(\alpha, \beta) = b^{-2\omega_c d} \sum_{i \in \alpha, j \in \beta} \left\langle S_i^{(n)} S_j^{(n)} \right\rangle. \tag{10.5}$$

A Primer to the Theory of Critical Phenomena. http://dx.doi.org/10.1016/B978-0-12-804685-2.00010-3

Provided that the blocks α, β are far apart, all correlations $\left\langle S_i^{(n)} S_j^{(n)} \right\rangle$ can be considered to be identical. We can then again replace the summation by b^{2d} identical terms, so that, approximately,

$$G^{(n+1)}(\alpha, \beta) = b^{2d(1-\omega_c)} G^{(n)}(i, j). \tag{10.6}$$

We now briefly anticipate the discussion of Chapter 11 by switching to a continuum representation. Let sites α and β be separated by the distance r. Then, for a translationally invariant system, it is appropriate to replace Eq. (10.6) by

$$G^{(n+1)}(br) = b^{2d(1-\omega_c)} G^{(n)}(r) \equiv b^{2d_\phi} G^{(n)}(r). \tag{10.7}$$

The factor b^{2d_ϕ} on the right provides the conventional interrelation between $G^{(n+1)}(br)$ and $G^{(n)}(r)$; further uses of the factor b^{2d_ϕ} are introduced later.

We next apply an important relation: as shown later, at the critical point, the distance dependence of $G(r)$ is empirically expressed in the form

$$G(r) = \frac{D}{r^{d-2+\eta}}, \tag{10.8}$$

where η is a small correction to $d-2$. In view of Eq. (10.7) and the redefinitions $r_2 = br$, $r_1 = r$, we find that

$$\frac{G(r_2)}{G(r_1)} = \left(\frac{r_2}{r_1}\right)^{d-2+\eta} = b^{d-2+\eta} = b^{2d(1-\omega_c)}. \tag{10.9}$$

Hence,

$$2\omega_c d = d + 2 - \eta, \tag{10.10}$$

which will be later used to eliminate ω_c.

10.2 EFFECT OF APPLIED MAGNETIC FIELDS

As before, we must be careful in dealing with alterations of the magnetic properties during Kadanoff rescaling, where the external magnetic field B acts on a continually decreasing number of involved spins. This cannot represent the actual physical process in which B interacts with the entire lattice. We therefore replace B with an effective field $B^{(n)}$ at the nth iteration of the renormalization process. Then the Hamiltonian $\hat{\mathcal{H}}^{(n)}$ contains the magnetic contribution $\mu B^{(n)} \sum_i S_i^{(n)}$, ($\mu \equiv g\mu_B$). To keep the Hamiltonian invariant at the next iteration, we must use $\mu B^{(n+1)} \sum_\alpha S_\alpha^{(n+1)}$ as the appropriate term in $\hat{\mathcal{H}}^{(n+1)}$. In view of Eq. (10.1), we then find that $\hat{\mathcal{H}}^{(n+1)}$ involves

$$\mu B^{(n+1)} \sum_{\alpha} S_{\alpha}^{(n+1)} = \mu B^{(n+1)} b^{-\omega_c d} \sum_{i} S_i^{(n)} = \mu B^{(n)} \sum_{i} S_i^{(n)}, \tag{10.11}$$

which, for a fixed field B, maintains the proper form invariance of $\hat{\mathcal{H}}$ during the changing spin number, provided that we set

$$B^{(n+1)} = b^{\omega_c d} B^{(n)}. \tag{10.12}$$

As introduced in Chapter 3, Eq. (3.9), the magnetic response at the critical point to a small variable magnetic field is expressed by the relation $M = \mu B^{1/\delta}$. For the progression from stage n to stage $n+1$, we then find that

$$\frac{M^{(n+1)}}{M^{(n)}} = \left(\frac{B^{(n+1)}}{B^{(n)}} \right)^{1/\delta}. \tag{10.13}$$

Now introduce Eqs. (10.3) and (10.12), whereby

$$b^{d(1-\omega_c)} = b^{\omega_c d/\delta}, \tag{10.14}$$

whence, according to Eq. (10.10),

$$\delta = \frac{\omega_c}{1-\omega_c} = \frac{d+2-\eta}{d-2+\eta}. \tag{10.15}$$

This relation fixes the value of ω_c if other exponents are determined from elsewhere.

10.3 MAGNETIZATION CLOSE TO THE CRITICAL TEMPERATURE

Close to the critical temperature, the magnetization is conventionally expressed as $M \sim \frac{|T-T_c|^{\beta}}{T_c}$. On introducing the correlation length $\xi = \frac{|T-T_c|^{-\nu}}{T_c}$, we note that $M \sim \xi^{-\beta/\nu}$. At two different stages of the blocking progress,

$$\frac{M^{(n+1)}}{M^{(n)}} = \left(\frac{\xi^{(n+1)}}{\xi^{(n)}} \right)^{-\beta/\nu}. \tag{10.16}$$

Now introduce Eq. (10.3) on the left and the correlation lengths before and after rescaling $\xi^{(n+1)} = \frac{\xi^{(n)}}{b}$ on the right. Then

$$b^{d(1-\omega_c)} = b^{\frac{\beta}{\nu}}, \tag{10.17}$$

so that, by Eqs. (10.10) and (10.15),

$$\beta = \nu d(1 - \omega_c) = \frac{\nu(d - 2 + \eta)}{2}. \tag{10.18}$$

Again, this relation fixes ω_c if the other components are known.

10.4 MAGNETIC SUSCEPTIBILITY NEAR THE CRITICAL TEMPERATURE

In light of our earlier discussion involving (10.1) and (10.12), we can express the magnetic susceptibility in terms of $M^{(n)}$ as

$$\chi^{(n)} = \frac{N\mu M^{(n)}}{B^{(n)}}. \tag{10.19}$$

Using Eqs. (10.16), (10.17), and (10.12), we examine the ratio of susceptibilities at two successive stages of renormalization at a constant applied external field:

$$\frac{\chi^{(n+1)}}{\chi^{(n)}} = \frac{B^{(n)}}{B^{(n+1)}} \frac{M^{(n+1)}}{M^{(n)}} = b^{-\omega_c d} b^{d(1-\omega_c)} = b^{d(1-2\omega_c)}. \tag{10.20}$$

Conventionally, the divergence of the magnetic susceptibility with temperature near the critical point is expressed through the relation

$$\chi \sim |T - T_c|^{-\gamma}. \tag{10.21}$$

On again using $\xi = |T - T_c|^{-\nu}$, we observe that $\chi \sim \xi^{\frac{\gamma}{\nu}}$, so that

$$\frac{\chi^{(n+1)}}{\chi^{(n)}} = \left(\frac{\xi^{(n+1)}}{\xi^{(n)}}\right)^{\frac{\gamma}{\nu}} = b^{-\frac{\gamma}{\nu}}. \tag{10.22}$$

Then, comparing (10.20) with (10.10) and (10.22), we find that

$$\gamma = -\nu d(1 - 2\omega_c) = \nu(2 - \eta). \tag{10.23}$$

10.5 HEAT CAPACITY

We first provide a heuristic explanation: As is made plausible in Chapter 5, the energy change in a process executed close to the critical point is of order $\Delta E \sim |t|^{\nu d}$. The heat capacity density is conventionally represented by the relation $c = \left|\frac{T-T_c}{T_c}\right|^{-\alpha}$, which translates to a free energy density in the form $f \sim |t|^{2-\alpha}$. On simply equating the two expressions, we find that

$$\alpha = 2 - \nu d, \tag{10.24}$$

which is *Josephson's law*. We have obviously left several important questions unresolved.

The more rigorous derivation of Josephson's law requires more work: we follow the procedure by Le Bellac.[1] By way of review of earlier work in Chapter 9, so far we have heavily depended on Eq. (10.1) in relating spin variables at stage n to those at stage $n+1$ during the renormalization process. This is unduly restrictive. Generalizing, we proceed as in Chapter 9. Namely, we relate the Hamiltonians at stages n and $n+1$ in the following generic manner:

$$\exp\left[-\hat{\mathcal{H}}^{(n+1)}\left(\left\{S_\alpha^{(n+1)}\right\}\right)\right] = \exp[-G]\cdot\sum\exp\left[-\hat{\mathcal{H}}^{(n)}\left(\left\{S_i^{(n)}\right\}\right)\right], \tag{10.25}$$

where the summation extends over all sets $S_i^{(n)}$ consistent with the sets $\left\{S_\alpha^{(n+1)}\right\}$. For present purposes, it is not necessary to specify the function G, except to note that it does not involve the spin variable set $\left\{S_\alpha^{(n+1)}\right\}$.

Now sum Eq. (10.25) over all $S_\alpha^{(n+1)}$ configurations and take logarithms to obtain the corresponding free energies ($\beta = 1$):

$$F^{(n+1)}\left[\hat{\mathcal{H}}^{(n+1)}\left(\left\{S_\alpha^{(n+1)}\right\}\right)\right] = F^{(n)}\left[\hat{\mathcal{H}}^{(n)}\left(\left\{S_i^{(n)}\right\}\right)\right] - G\left[\hat{\mathcal{H}}^{(n)}\left(\left\{S_i^{(n)}\right\}\right)\right]. \tag{10.26}$$

As before, in place of the Hamiltonians, we can equally use the collection of parameters K_i combined into components of the vector $\boldsymbol{\kappa}$. We now introduce intensive variables via $F^{(n)}(\boldsymbol{\kappa}^{(n)}) = N^{(n)} f^{(n)}(\boldsymbol{\kappa}^{(n)})$. Similarly, $F^{(n+1)}(\boldsymbol{\kappa}^{(n+1)}) = N^{(n+1)} f^{(n)}(\boldsymbol{\kappa}^{(n)})$; note that we have again used the same function $f^{(n)}$ [Why?]. Lastly, we write $G = N^{(n)} g(\boldsymbol{\kappa}^{(n)})$. With minor changes in notation, this leads us to

$$f(\boldsymbol{\kappa}_0) = g(\boldsymbol{\kappa}_0) + b^{-d} f(\boldsymbol{\kappa}_1). \tag{10.27}$$

Now introduce our rescaling step over and over:

$$\begin{aligned} f(\boldsymbol{\kappa}_1) &= g(\boldsymbol{\kappa}_1) + b^{-d} f(\boldsymbol{\kappa}_2), \\ f(\boldsymbol{\kappa}_2) &= g(\boldsymbol{\kappa}_2) + b^{-d} f(\boldsymbol{\kappa}_3), \\ &\cdots\cdots\cdots\cdots\cdots\cdots \\ f(\boldsymbol{\kappa}_n) &= g(\boldsymbol{\kappa}_n) + b^{-d} f(\boldsymbol{\kappa}_{n+1}). \end{aligned} \tag{10.28}$$

1. M. Le Bellac, *Quantum and Statistical Field Theory* (Clarendon Press, Oxford, 1991), pp. 84 ff.

Then multiply the second equation by b^{-d}, ..., the nth equation by $b^{-d(n-1)}$, and add the resultant to obtain

$$f(\boldsymbol{\kappa}_0) = \sum_{n=0}^{\infty} b^{-nd} g(\boldsymbol{\kappa}_n). \tag{10.29}$$

Proceeding to the continuum limit, we convert the latter to

$$f(\boldsymbol{\kappa}_0) = \int d\boldsymbol{\kappa} b^{-nd} g(\boldsymbol{\kappa}(b)). \tag{10.30}$$

We had earlier introduced the relation $b = \left|\frac{T-T_c}{T_c}\right|^{1/y_1}$. In this range, close to T_c, $g(\boldsymbol{\kappa}(b))$ varies very slowly with b and may be regarded as a constant. After its placement outside, the integral converges to b^{-d}, whence, with $1/y_1 \equiv \nu$,

$$f(\boldsymbol{\kappa}_0) \sim b^{-d} \left|\frac{T - T_c}{T_c}\right|^{\nu d}. \tag{10.31}$$

We are almost there. As is well established, the heat capacity (density) is obtained from the relation $c = -\frac{\partial^2 f}{\partial T^2} = \left|\frac{T-T_c}{T_c}\right|^{\nu d-2}$. Conventionally, the heat capacity is written out as $c \sim \left|\frac{T-T_c}{T_c}\right|^{-\alpha}$. Thus,

$$\alpha = 2 - \nu d, \tag{10.32}$$

which is the Josephson equation.

10.6 SUMMARY

Let us summarize our discussion: the four critical exponents α, β, γ, and δ can all be expressed in terms of ν and η, with d as a parameter.

Summary of Interrelations:

- α (Josephson's law):

$$c \sim \left|\frac{T - T_c}{T_c}\right|^{-\alpha}, \quad \alpha = 2 - \nu d;$$

- β:

$$M \sim \left(\frac{T_c - T}{T_c}\right)^{\beta}, \quad \beta = \frac{\nu(d - 2 + \eta)}{2}; \tag{10.33}$$

- γ (Fisher's law):

$$\chi \sim \left|\frac{T - T_c}{T_c}\right|^{-\gamma}, \quad \gamma = \nu(2-\eta);$$

- δ:

$$M \sim B^{1/\delta}, \quad \delta = \frac{d+2-\eta}{d-2+\eta}.$$

By purely algebraic methods we readily obtain interrelations among the critical exponents, which involve neither ν nor η. Among the simplest are the following.

Selected Interrelations among Critical Exponents

- Rushbrooke's laws:

$$\begin{aligned} 2\beta + \gamma &= 2 - \alpha, \\ 2\beta\delta - \gamma &= 2 - \alpha. \end{aligned} \tag{10.34}$$

- Griffith's laws:

$$\begin{aligned} \alpha + \beta(1+\delta) &= 2, \\ \beta(1-\delta) + \gamma &= 0. \end{aligned}$$

All these interrelations have been carefully checked experimentally.

Chapter 11

Extension of the Landau Approach to Inhomogeneous Systems and the Physical Picture of Gaussian Fluctuations

Up to this point, we used the discrete lattice array of Ising spins (except in Chapter 4.3), on which we have built the block procedure used in the analysis of critical phenomena. We now seek to undertake a similar effort, based on the continuum representation. This change does not present a problem in the long-wave perturbation limit, where the position coordinate can always be made to satisfy the relation $r \gg a$, with a of the order of atomic dimensions, and with no upper limit on r. However, in dealing with phase transitions, the correlation distance ξ appears as an additional variable; in the scaling-up processes, the condition $a \ll r \ll \xi$ will have to be met. The way to enforce this is to place restrictions on the changes in interaction parameters u and λ of the Landau expansion at each stage of the rescaling, as will later be demonstrated. Among other items, this depends on the dimension d of the system; in light of our prior experience, we will often work instead with the parameter $\epsilon = 4 - d$. Here we discuss mainly the case of a one-component order parameter $\eta = M(\boldsymbol{r})$ (except Sections 11.3 and 11.4). This chapter directly corresponds to Chapter 4.

11.1 SMOOTHING OPERATIONS

First, we need to replace the discretized order parameter η with its corresponding order parameter, henceforth designated as the function $\phi(\boldsymbol{r})$ that varies continuously with position $\boldsymbol{r}$; other physical parameters, such as temperature or the magnetic field, will also be allowed to change continuously with $\boldsymbol{r}$. Here we follow the line of reasoning by Goldenfeld.[1]

We begin with the collection of spins in a d-dimensional hypercubic lattice, where close to criticality the lattice is organized into patches of size roughly

1. N. Goldenfeld, *Lectures on Phase Transitions and the Renormalization Group*, Perseus Books, Reading, MA, 1992, Chapter 5.

A Primer to the Theory of Critical Phenomena. http://dx.doi.org/10.1016/B978-0-12-804685-2.00011-5

corresponding to the correlation length $\xi(T)$. Within a patch (or domain), the magnetization (density) $M(\boldsymbol{r})$ varies only slightly. Correspondingly, we introduce a real-space cutoff parameter $\Lambda^{-1} \approx \xi(T) \gg a$, where a is the lattice parameter. Events at distances smaller than Λ^{-1} are replaced by an average local magnetization

$$M(\boldsymbol{r}) \equiv \frac{1}{N(\boldsymbol{r})} \sum_{i \in \boldsymbol{r}} S_i, \tag{11.1}$$

where $N(\boldsymbol{r}) = \Lambda^{-d}/a^d$ is the number of spins in the block centered on position $\boldsymbol{r}$. This coarse-grained magnetization $M(\boldsymbol{r})$ is to vary smoothly with position within each patch. Stated differently, events on the scale $r < \Lambda^{-1}$ are neglected, which means that the inhomogeneities (fluctuations) on the short-range scale are ignored (at energies $\gg k_B T_s$). With Λ^{-1} as the smallest distance of concern, we monitor progress along each axis in units of Λ^{-1}.

On this basis, we begin by changing the Landau functionals of Chapter 4 (particularly Eq. (4.5)) from a power series that involve η to the format

$$\mathcal{L} = \sum_{\boldsymbol{r}} \mathcal{L}\{M(\boldsymbol{r})\}, \tag{11.2}$$

where we sum over all patches; $\mathcal{L}\{M(\boldsymbol{r})\}$ is now to be represented by a power series. However, we also need to include an additional requirement: in proceeding from patch to patch, the change in $M(\boldsymbol{r})$ should be minimal. For, large differences in the properties of contiguous blocks that set up domain walls, are energetically very unfavorable. The simplest way to avoid, or at least minimize, their existence is to introduce a term of the type

$$\nabla M(\boldsymbol{r}) \equiv \lim_{\Lambda^{-1} \to 0} \frac{1}{2}\alpha \left[\frac{M(\boldsymbol{r}) - M(\boldsymbol{r} + \delta \boldsymbol{r})}{\Lambda^{-1}}\right]^2 \tag{11.3}$$

with $\alpha > 0$, which are supposed to change smoothly and incrementally as one proceeds from patch to patch. The factor $1/2$ is added for later convenience, $\alpha \sim J$ is the interaction parameter, and $\delta \boldsymbol{r}$ is a vector with magnitude Λ^{-1} that points to an adjacent block. The coefficient α may vary with temperature. On the assumption that $M(\boldsymbol{r})$ changes relatively slowly with position, we may then replace the bracketed term in the summation (11.3) by the square of the gradient vector, as was done in Sec. 4.3. Moreover, the summations in Eqs. (11.2) and (11.3) may be replaced by integrations. When these contributions are introduced, we obtain

$$\mathcal{L} = \int d^d \boldsymbol{r} \left\{\mathcal{L}\{M(\boldsymbol{r})\} + \frac{1}{2}\alpha \left[\nabla M(\boldsymbol{r})\right]^2\right\}. \tag{11.4}$$

Since the block structure suppresses fluctuations within the block of average size ξ, the scalar order parameter $\phi(\boldsymbol{r}) \equiv M(\boldsymbol{r})$ is expected to vary smoothly in proceeding from one block to the next. This, in turn hopefully, renders the square of the gradient sufficient to suppress large local variations. However, this advantage comes at a price: as long as $\alpha = 0$, that is, in the absence of the gradient term, $\phi(\boldsymbol{r})$ is a purely local quantity. When the term (11.3) is incorporated into the Landau formalism, the Landau free energy functional depends on what happens elsewhere in the system; thus, the strict locality is abandoned. The expression $\mathcal{L}$ represents an example of a functional of $M(\boldsymbol{r})$, which means that the ordinary minimization of $\mathcal{L}$ with respect to $M(\boldsymbol{r}) = \text{const} \equiv \eta$ from Chapter 4, is replaced by the corresponding variational procedure. However, we will go next beyond the mean-field approximation. This reasoning represents so far an intuitive justification of the formal approach formulated in the second half of that in the Chapter 4. We also introduce a new aspect, namely the thermodynamic fluctuations of the order parameter at thermal equilibrium when approaching the critical point.

11.2 USE OF FUNCTIONAL INTEGRALS

To summarize, we have replaced the summations appropriate to discrete values of the order parameter of Chapter 4 by a functional integral starting from the functional $\mathcal{L} = \mathcal{L}(\{\phi(\boldsymbol{r})\})$ in slightly nonstandard form, that is,

$$\mathcal{L} = \int d^d\boldsymbol{r} \left\{ \frac{1}{2}\alpha[\nabla\phi(\boldsymbol{r})]^2 + \frac{1}{2}ut\,[\phi(\boldsymbol{r})]^2 + \frac{\lambda}{4!}[\phi(\boldsymbol{r})]^4 - B\phi(\boldsymbol{r}) \right\} \tag{11.5}$$

with $t \equiv (T - T_c)/T_c$ as before. Here we have rewritten the coefficients of Chapter 4 so as to be in conformity with standard notation.[2] The corresponding partition function involves the Hamiltonian density function $\mathcal{L}(\{\phi(\boldsymbol{r})\})$:

$$\begin{aligned} \mathcal{Z} &= \int \mathcal{D}\phi \exp\left\{-\beta \int d^d\boldsymbol{r}\, \mathcal{L}[\phi(\boldsymbol{r})]\right\} \\ &= \int \mathcal{D}\phi(\boldsymbol{r}) \exp\left\{-\beta \int d^d r \left[\frac{1}{2}\alpha(\nabla\phi(\boldsymbol{r}))^2 + \frac{1}{2}ut\,[\phi(\boldsymbol{r})]^2 \right.\right. \\ &\quad \left.\left. + \frac{\lambda}{4!}[\phi(\boldsymbol{r})]^4 - B\phi(\boldsymbol{r})\right]\right\}. \end{aligned} \tag{11.6}$$

The integration measure $\mathcal{D}\phi$ includes a summation over all functions $\{\phi(\boldsymbol{r})\}$ reflecting all possible configurations of the field $\phi(\boldsymbol{r})$. Later, we will specify

2. The functional (11.5), with the gradient term present, is customarily called the Ginzburg–Landau free energy functional. It was introduced in the context of the theory of superconductivity (1950), later recognized as of universal validity in classical continuous phase transitions. Note also that $\mathcal{L} \equiv \mathcal{F} - \mathcal{F}_0$ (cf. Chapter 4).

this functional integration $\int \mathcal{D}\phi$ explicitly for the particular case of Gaussian fluctuations.

To review again the physical significance of the various coefficients, α describes the interactions between adjacent regions at given positions, u is a parameter that introduces the thermal characteristic fluctuations involved in the details of a phase transition when T crosses T_c, as discussed in Chapter 4; this contribution includes the linear scaled temperature t. $\lambda > 0$ (negative values would allow for the possibility that the resulting probabilities become unbounded) allows for nonlinear (non-Gaussian) effects. Similar to the first term, it suppresses large order parameter fluctuations. B plays a dual role: aside from specifying the energy in an applied magnetic field, it also allows us to derive formally the correlation functions by differentiating $\mathcal{Z}$, as detailed in Chapter 3. The various factors of $1/2$ and 4! are included for convenience; we again set $\beta = \frac{1}{k_B T} = 1$; it is supposed that only the difference $T - T_c$ matters in the critical regime.

Interpretation of the Formalism

In accordance with the general philosophy of the Landau theory of phase transitions,[3] expression (11.5) represents the work required to create a fluctuation in the system having the fluctuation space profile $\phi(\boldsymbol{r})$. Therefore, the functional integration amounts to summing up all possible (classical) fluctuation profiles that may appear in the system at a given temperature T. In these circumstances the fluctuations in equilibrium are regarded as the degrees of freedom, in addition to the reference (mean-field) value of the order parameter $M(\boldsymbol{r}) \equiv \text{const} = \eta$. These are obtained from the minimization of the Landau functional discussed in Chapter 4, where **all** thermodynamic fluctuations are disregarded, so that the order parameter acquires a single value obtained from the minimization of $\mathcal{F}$. Such a minimization, when carried out for (11.5), would provide a single (frozen) space profile $\phi(\boldsymbol{r})$, which appears when considering, for example, (mean-field) variations of $M \equiv \phi(\boldsymbol{r})$ near surfaces, as in thin films, and so on. Only the integration (11.6) provides a full thermodynamic description of a continuous phase transition. In effect, we have to introduce the physical (renormalized) free energy $\mathcal{F}(T, B)$ of the system in which the fluctuations provide a prominent (or even singular) contribution. This takes the form

$$\mathcal{F}(T, B) = -k_B T \ln \mathcal{Z}(T, B) + \mathcal{F}_0. \tag{11.7}$$

3. L.D. Landau and E.M. Lifshitz, *Statistical Physics*, 3rd edition, Part 1 (Elsevier, Amsterdam, 1980), §147.

11.3 ROCK-BOTTOM APPROXIMATION

It may be helpful to show how the preceding reduces to the standard Landau formulation. For this purpose, we replace the partition function (11.6), by its largest term.[4] To achieve this, in the functional appearing in the exponent, we set $\nabla\phi(\boldsymbol{r}) = 0$, whence ϕ is constant (as $V \to \infty$), its stationary value. Finally, we discard the term involving B. With these simplifications, we find that (V is the volume)

$$\int d^d\boldsymbol{r}\, \mathcal{L}[\phi(\boldsymbol{r})] = V\left(\frac{1}{2}ut\,\phi^2 + \frac{\lambda}{4!}\phi^4\right), \tag{11.8}$$

which directly leads to the Landau free energy density of Chapter 4,

$$\frac{\mathcal{F}}{V} = \mathcal{F}_0 + \frac{1}{2}ut\phi^2 + \frac{\lambda}{4!}\phi^4. \tag{11.9}$$

This is clearly an extension of the Landau theory (from the case $\phi = \text{const} \equiv \eta$). However, near the critical temperature ($\phi \to 0$), the order parameter can fluctuate wildly; thus, the approximation involving a constant ϕ is obviously inadequate. The principal question is how close to $T = T_c$ we have to come so that the fluctuations represent the dominant contribution.

11.4 FOURIER TRANSFORMS AND SCALING IN RECIPROCAL SPACE

Before launching into the next stage, it is convenient to shift operations to the *reciprocal* or ***k*** *space* via the Fourier transform of a function $f(\boldsymbol{x})$ in a d-dimensional space,[5]

$$f(\boldsymbol{k}) = \int_{-\infty}^{\infty} d^d\boldsymbol{x} e^{-i\boldsymbol{k}\cdot\boldsymbol{r}} f(\boldsymbol{x}), \tag{11.10}$$

together with its reciprocal relation

$$f(\boldsymbol{r}) = \int_{-\infty}^{\infty} \frac{d^d\boldsymbol{k}}{(2\pi)^d}\, e^{i\boldsymbol{k}\cdot\boldsymbol{r}} f(\boldsymbol{k}). \tag{11.11}$$

---

4. This crude approximation is called *the saddle-point approximation.* It will be shown later that it corresponds to the standard mean-field approximation (MFA). The present starting point has the principal advantage over the standard MFA discussed earlier that we can study fluctuations with respect to this reference (broken-symmetry) state by considering deviations $\delta\phi \equiv \phi(\boldsymbol{r}) - \phi$.

5. This formal trick has deep physical roots: We no longer describe fluctuations in terms of real-space variations $\delta\phi \equiv \phi(\boldsymbol{r}) - \phi$, but by the wavelength $\lambda = 2\pi/k$ of the periodic Fourier components, their equivalent to the position in the system. In effect, this transformation into the reciprocal ($\boldsymbol{k}$) language enormously simplifies the formal (and physical) discussion.

The introduction of this transform requires that we become familiar with several operations in reciprocal ($\boldsymbol{k}$) space. Namely, earlier we encountered rescaling procedures by which we retained only distances greater than some cutoff value: thus, $x \geq \Lambda^{-1}$. The corresponding restriction in reciprocal space reads: $k \leq \Lambda$. The length change $x = bx'$ corresponds to $k = \frac{k'}{b}$ in reciprocal space. The direct space integration

$$\int d^d \boldsymbol{x}' = b^{-d} \int d^d \boldsymbol{x} \tag{11.12}$$

corresponds to the reciprocal space operation

$$\int d^d \boldsymbol{k}' = b^{d} \int d^d \boldsymbol{k}. \tag{11.13}$$

Note that the factor $\boldsymbol{k} \cdot \boldsymbol{r}$ remains invariant under these operations. This operation means that an elongation by b in direct space is matched by a corresponding dilatation by $1/b$ in k space, that is, shifting the fluctuating Fourier components to the longer-wavelength regime. The RGT operations hinge on a rescaling of the distance x along a single direction via $\Delta x' = b^{-1}\Delta x \equiv b^{-d_s}\Delta x$. It is customary to introduce the concept of *scale dimension* d_s, defined by the exponent of the scale factor b; here $d_s = -1$, a counterintuitive value. More generally, given a relation $f' = b^n f$ between two functions or variables, we have $d_s = n$. Unfortunately, several authors do not adequately distinguish between scale dimensions and ordinary dimensions; so, be on guard when reading the literature.

A particular rescaling of interest involves the order parameter functions $\phi(x)$. It is conventional to renormalize such fields according to the equation $\phi(x) = b^{-d_\phi}\phi'(x')$, whence the Fourier transform operation involves the relation

$$\phi'(\boldsymbol{k}') = \int_{-\infty}^{\infty} d^d \boldsymbol{r}' e^{-i\boldsymbol{k}'\cdot\boldsymbol{r}'}\phi'(\boldsymbol{r}') = b^{(d_\phi - d)} \int_{-\infty}^{\infty} d^d \boldsymbol{r} e^{-i\boldsymbol{k}\cdot\boldsymbol{r}}\phi(\boldsymbol{r}) = b^{d_\phi - d}\phi(\boldsymbol{k}), \tag{11.14}$$

where d_ϕ is known as the *anomalous dimension* of the field (order parameter).

As an aside, correlation functions involve the product of two order parameters $\phi(x)$. Accordingly, $G'(r') = b^{2d_\phi}G(r)$. At a fixed point, $G^*(r/b) = b^{2d_\phi}G^*(r)$. Then, with $G^*(r) = b^{-2d_\phi}G^*(1) \sim 1/r^{(d-2+\eta)}$ (n.b., here η *is not* the order parameter previously used), it is clear that the anomalous dimension is given by

$$d_\phi = \frac{(d-2+\eta)}{2}. \tag{11.15}$$

11.5 EXTENSION: THE GINSBURG–LANDAU FUNCTIONAL IN A ROTATIONALLY INVARIANT CASE

The above discussion involved a single-component order parameter $\phi(\boldsymbol{r})$. However, all this can be generalized at this level to the rotationally invariant form $\phi(\boldsymbol{r}) \to \boldsymbol{M}(\boldsymbol{r}) = \{M_i(\boldsymbol{r})\}_{i=1,2,3}$ in a d-dimensional space. For that purpose, we return to expression (4.91) (or equivalently, to (4.93)), which can be rewritten in the form

$$\mathcal{L} \equiv \int d^d\boldsymbol{r} \left\{ \frac{1}{2}\alpha|\nabla \boldsymbol{M}(\boldsymbol{r})|^2 + \frac{1}{2}ut|\boldsymbol{M}(\boldsymbol{r})|^2 + \frac{\lambda}{4!}|\boldsymbol{M}(\boldsymbol{r})|^4 - \boldsymbol{B}\cdot\boldsymbol{M}(\boldsymbol{r}) \right\}. \tag{11.16}$$

The three degrees of freedom of $\boldsymbol{M}$ corresponding to magnetization components become inequivalent for the case of an anisotropic paramagnet undergoing a para- to ferromagnetic phase transition. Then the integration over the profiles of $\boldsymbol{M}(\boldsymbol{r})$ involves an integration over both angular and amplitude ($|\boldsymbol{M}(\boldsymbol{r})|$) thermodynamic fluctuations. However, these two types of fluctuations become inequivalent in the ferromagnetic phase since when $T < T_c$, we encounter a spontaneous breakdown of the rotational symmetry where a static magnetization η appears of the form $\boldsymbol{M}(\boldsymbol{r}) = \eta(T, B)\hat{e}_z + \boldsymbol{m}(\boldsymbol{r})B$. In the ordered phase, $\eta \neq 0$, $(T < T_c)$; then, with decreasing temperature, the fluctuation component $m_z(\boldsymbol{r})$ reduces gradually to zero. For $0 < T \ll T_c$, only the transverse thermal fluctuations are relevant; their simplest form are the spin waves, that is, the periodic waves representing elementary collective excitations. However, we are interested in the complementary regime $T \to T_c - 0$ and must therefore distinguish between transverse fluctuations, involving the components (m_x, m_y), and longitudinal (m_z) fluctuations, which are called the *amplitude-mode fluctuations*. Above T_c, these two types of fluctuations coalesce into three equivalent channels.

In the magnetic case discussed here, we can represent the transverse fluctuations as circular components $m^{\pm} \equiv m_x \pm i m_y$, which represent the equivalence of those components in the presence of longitudinal order. The character of those fluctuations reflects the fact that the magnetic moment is antialigned with the spin angular moment, which in turn undergoes a precessional motion under the influence of the magnetic torque $\boldsymbol{M} \times \boldsymbol{B}$, wherc $\boldsymbol{B}$ involves either an external $(\boldsymbol{H}_a)$ or a local $(\boldsymbol{H}_i(\boldsymbol{r}))$ field, or the sum of both. In general, we can thus define the collective excitations of such a system via transverse $\chi^{+-} \sim \langle m^+(\boldsymbol{r}, t)\, m^-(\boldsymbol{r}', t')\rangle$ and longitudinal $\chi^{zz} \sim \langle m^z(\boldsymbol{r}, t)\, m^z(\boldsymbol{r}', t')\rangle$ correlation functions, which represent generalized magnetic susceptibilities. These susceptibilities describe general fluctuation modes, phase and amplitude components, as drawn schematically in Fig. 11.1. These considerations are elaborated on in the next chapters.

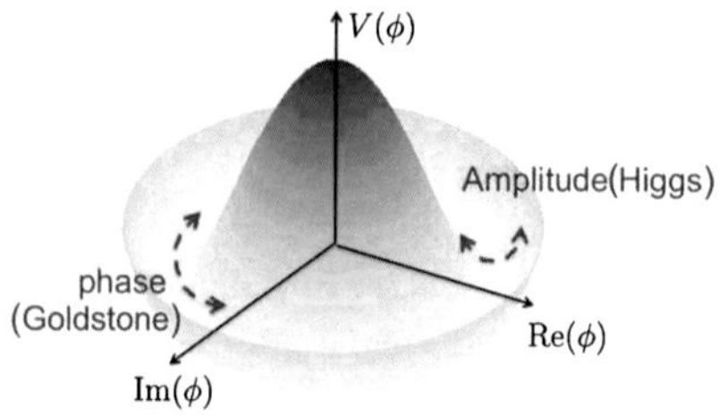

FIGURE 11.1 Schematic representation of the part of the Ginzburg–Landau functional without the gradient term (the effective potential) $V(\phi)$ for a two-component order parameter ($d = 2$) with phase ("Goldstone" mode) and amplitude ("Anderson–Higgs" mode) fluctuations. This takes place in the broken-symmetry (ordered) phase.

11.6 A CONCRETE EXAMPLE: GAUSSIAN FLUCTUATIONS AND THE GINZBURG CRITERION

Consider a three-dimensional system with a local scalar order parameter $\eta(\boldsymbol{r})$ with $T \geqslant T_c$. We next introduce the so-called *Gaussian approximation*,[6] that is, take the functional (11.5) for $B = 0$ in the form

$$\mathcal{L} = \int d^3r \left\{\alpha\, t\, \eta^2(\boldsymbol{r}) + g(\nabla\eta)^2\right\}. \tag{11.17}$$

This functional does not contain the essential term $\sim \eta^4$ required to obtain a stable nonzero mean-field value of η. According to our prescription, we take the Fourier transform $\eta(\boldsymbol{r})$, that is,

$$\eta(\boldsymbol{r}) \equiv \sum_{\boldsymbol{k}} \eta_{\boldsymbol{k}}\, e^{i\boldsymbol{k}\cdot\boldsymbol{r}}. \tag{11.18}$$

Note that since the order parameter is a real-valued function, $\eta(\boldsymbol{r}) \equiv \eta^*(\boldsymbol{r})$, we must require that $\eta_{\boldsymbol{k}} = \eta^*_{-\boldsymbol{k}}$. Analogously,

$$\nabla\eta(\boldsymbol{r}) = \sum_{\boldsymbol{k}} i\,\boldsymbol{k}\,\eta_{\boldsymbol{k}}\, e^{i\boldsymbol{k}\cdot\boldsymbol{r}}. \tag{11.19}$$

In fact, the functional can be brought to a closed form

$$\mathcal{L}\{\eta_{\boldsymbol{k}}\} = V \sum_{\boldsymbol{k}} \left(\alpha t + g k^2\right) |\eta_{\boldsymbol{k}}|^2 \tag{11.20}$$

6. After L.D. Landau and E.M. Lifshitz, *Statistical Physics*, 3rd edition, Part 1 (Elsevier, Amsterdam, 1980), p. 482.

since $\eta_{-\boldsymbol{k}} = \eta^*_{\boldsymbol{k}}$. Thus, the physical partition function (11.6) including the fluctuations takes the form

$$\mathcal{Z} = \int \mathcal{D}\eta(\boldsymbol{r})\, e^{-B\mathcal{L}\{\eta(\boldsymbol{r})\}} = \prod_{\boldsymbol{k}} d\eta'_{\boldsymbol{k}}\, d\eta''_{\boldsymbol{k}}\, e^{-\beta\mathcal{L}\{\eta(\boldsymbol{k})\}}, \tag{11.21}$$

where the integration is performed over both real ($\eta'_{\boldsymbol{k}}$) and imaginary ($\eta''_{\boldsymbol{k}}$) parts. Note that $d^2\eta_k = |\eta_k| d|\eta_k| d\phi_k$. Also, in conformity with our earlier assumption, we introduce the cutoff in reciprocal space as $|\boldsymbol{k}| < \Lambda$. In fact, our physical free energy takes the form (up to a constant)

$$F(T) = F_0 - k_B T \sum_{|\boldsymbol{k}|<\Lambda} \ln \frac{\pi T}{\alpha t + g k^2} = F_0 - k_B T V \int_0^{\Lambda} \frac{k^2 dk}{4\pi^2} \ln\left(\frac{\alpha t + g k^2}{\pi T}\right). \tag{11.22}$$

This integration has been carried out over only half of $\boldsymbol{k}$-space since $\eta_{\boldsymbol{k}}$ and $\eta_{-\boldsymbol{k}}$ are not independent. Also, note that the functional integration over all possible space profiles $\{\eta(\boldsymbol{r})\}$, by going to the Fourier space, has been reduced to the product of Gaussian integrals over k-exponents, which is doable in an elementary manner. By carrying out a detailed integration and taking into account the relation that the specific heat (per unit volume) is $C(T) = -\left(\frac{T}{V}\right) \partial^2 F/\partial T^2$, we obtain for $T \to T_c$ that

$$C(T) - C_0 = \frac{k_B T_c \alpha^2}{4\pi^2} \int_0^{\infty} \frac{k^2 dk}{\left(\alpha t + g k^2\right)^2} = \frac{k_B T_c \alpha^{3/2}}{16\pi g^{3/2}} \frac{1}{\sqrt{t}}. \tag{11.23}$$

Setting $t \equiv (T - T_c)/T_c$, we see that the specific heat diverges with the critical exponent $\alpha = 1/2$. This result can be contrasted with the corresponding expression in the mean-field approximation, which increases linearly (for $T < T_c$) and exhibits a finite discontinuity. Namely, the mean field expression for the specific heat can be derived in an elementary manner. For $T \leqslant T_c$, according to Eq. (11.7) in the present notation, we have

$$\mathcal{L} = \alpha t \eta^2 + g \eta^4. \tag{11.24}$$

From $\partial\mathcal{L}\,\partial\eta = 0$ we obtain

$$\eta^2 \equiv \eta_0^2 = -\frac{\alpha t}{2g}; \quad F = F_0 - \frac{\alpha^2 t^2}{4g}. \tag{11.25}$$

The nontrivial part of the specific heat thus has the form

$$C^{MF}(T) - C_0 = \frac{\alpha^2}{2g} T \approx \frac{\alpha^2}{2g} T_c. \tag{11.26}$$

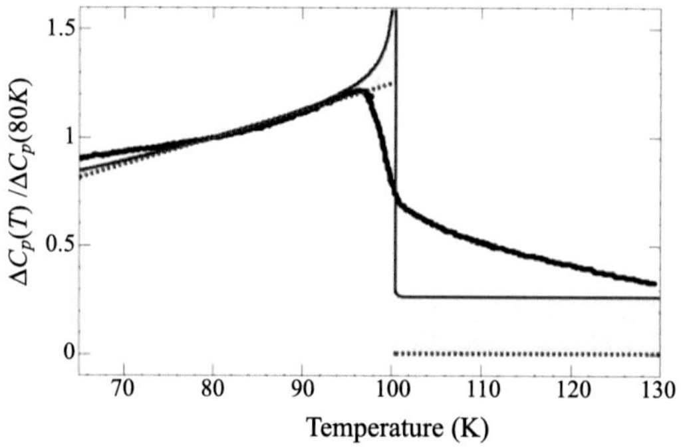

FIGURE 11.2 Temperature dependence of the specific heat for $BiMnO_3$. Dotted line: the mean-field part of the specific heat; thick solid line: experimental data were taken as $C_p|_{BiScO_3}$. Thin solid line: specific heat after taking into account the Gaussian thermal fluctuations (after O. Howczak and J. Spałek[7]).

In fact, we can roughly determine the temperature range ΔT in which the fluctuation contribution (11.23) exceeds the mean-field value (11.26); that is, for $T_c \geqslant T \geqslant T_c - \Delta T$,

$$C(T) \geqslant C^{MF}(T). \tag{11.27}$$

The equality sign in Eq. (11.27) defines the so-called *Ginzburg criterion* for the extension of the (classical) critical regime around T_c. A detailed analysis becomes more involved. Parenthetically, in the case of the Bardeen–Cooper–Schrieffer (BCS) theory of superconductivity the mean field is almost exact in describing the so-called type I superconductors, whereas in magnetic (Ising, Heisenberg) systems the role of the fluctuations beyond the Gaussian approximation is essential in determining the correct critical behavior (e.g., the critical exponents). The difference between those two systems is due to the fact that the BCS theory relies on a contact interaction in $\boldsymbol{k}$-space (hence, extends over large distances in real space), whereas the magnetic (exchange) involves short-range interactions (between the nearest neighbors, rarely also between the next-nearest neighbors) in the direct space. An essentially increased range of interaction between particles or other constituents in direct space may lead to the suppression of fluctuations. In fact, for interactions that extend equally to all spins or other objects (which have an infinite range), the mean field solution becomes exact. These issues are not discussed in this book.

A reader interested in more involved situations, with multiple component order parameters, may consult, for example, the paper of Howczak and Spałek (2010).[7] Note also that if the order parameter in (11.7) acquires D components

7. O. Howczak, J. Spałek, Eur. J. Phys. B **78**, 417–428 (2010).

(e.g., $\eta(\boldsymbol{r}) \rightarrow \boldsymbol{r}$), then the contributions to the specific heat (11.23) are multiplied by the factor D. The reader is asked to explain why and under what circumstances this is the case.

In Fig. 11.2, we provide a comparison of the mean-field theory with that incorporating the Gaussian fluctuations. The experimental data are taken for the ferromagnetic–ferroelectric system $BiMnO_3$ (quoted in footnote 7). The theory is based on two one-component ($D = 2$) order parameters: magnetization and ferroelectric dipole moment. Neither of the theoretical approaches properly interprets the data, even though they do not exhibit a clear singularity (λ-type transition). All this illustrates the fact that the transition from a qualitative to quantitative theoretical description of real experiments is not an easy task.

Chapter 12

The Ginzburg–Landau Functional for a Continuous System: Formal Approach to Gaussian Fluctuations

We finally start on our project of studying RGT operations based on the continuum approach. In this chapter, we put on a more formal basis the discussion of Chapter 11 and extend it by including the quartic term $\sim \lambda\phi^4$. In Appendices 12.A–12.D, we provide some details of our analysis.

12.1 GAUSSIAN INTEGRALS

We begin again with a truncated version of the partition function for the Ginzburg–Landau model, which is consistent with the Ginzburg–Landau free energy functional, Eq. (11.6), namely ($\beta = 1$)

$$\begin{aligned}\mathcal{Z}_G[J] = \int \mathcal{D}\phi \exp\left[-\int d^d\boldsymbol{x}\left(\frac{1}{2}\alpha\,|\nabla\phi|^2 + \frac{1}{2}u\phi^2 - J\phi\right)\right] \\ \equiv \int \mathcal{D}\phi \exp\left(-\hat{\mathcal{H}}_0[J]\right).\end{aligned} \tag{12.1}$$

Here $\phi(\boldsymbol{x})$ is the order parameter appropriate to the continuum, $\int \mathcal{D}\phi$ signifies a "functional integral" over all conceivable forms of the function $\phi(\boldsymbol{x})$ of relevance to the physical processes under investigation. The catalogue of such functions $\phi(\boldsymbol{x})$ excludes all forms that are not "well behaved," that is, vary wildly on the scale $\boldsymbol{r} \lesssim \xi$, since then the gradient term would take very large values. Also, as discussed in Chapter 11, from the set of acceptable functions we must exclude those portions that exhibit fluctuations on a scale of microscopic dimensions $x = |x| < \Lambda^{-1}$. The concept of functional integration and its execution are briefly addressed in Appendix 12.C.

In what follows, it is useful to work in reciprocal rather than in direct space. As is customary, we introduce the Fourier transforms that relate the position variable $\boldsymbol{x}$ to the wave vector $\boldsymbol{k}$, except that we must restrict $k = |\boldsymbol{k}|$ to the range

A Primer to the Theory of Critical Phenomena. http://dx.doi.org/10.1016/B978-0-12-804685-2.00012-7

$0 \leq k \leq \Lambda$, where Λ functions as a cut-off parameter in reciprocal space. Thus,

$$\phi(\boldsymbol{x}) = \frac{1}{(2\pi)^d} \int_{0 \leq k \leq \Lambda} d^d \boldsymbol{k} e^{i\boldsymbol{k}\cdot\boldsymbol{x}} \phi(\boldsymbol{k}). \tag{12.2}$$

With $\boldsymbol{k}$ as a reciprocal distance vector, $\phi(\boldsymbol{x})$ now varies smoothly over distances greater than Λ^{-1}, as required. These are the basics.

Before going further, we need to engage in an exercise: consider the following integral:

$$G[J] = \int \mathcal{D}\phi \exp\left\{ - \iint d^d \boldsymbol{x} d^d \boldsymbol{y} \phi(\boldsymbol{y}) \hat{\mathcal{M}}(\boldsymbol{x}, \boldsymbol{y}) \phi(\boldsymbol{x}) + \int d^d z \phi(z) J(z) \right\}, \tag{12.3}$$

where the operator $\hat{\mathcal{M}}(\boldsymbol{x}, \boldsymbol{y})$ is a symmetric function in the indicated variables that acts on $\phi(\boldsymbol{x})$.

Now introduce $\psi_n(\boldsymbol{x})$ as a set of orthonormal functions that serve as basis vectors for the space in which $\hat{\mathcal{M}}(\boldsymbol{x}, \boldsymbol{y})$ operates. Then any function in the Hilbert space spanned by these $\{\psi_n(\boldsymbol{x})\}$, in particular, $\phi(\boldsymbol{x})$ and $J(\boldsymbol{x})$, can be expanded in terms of such basis vectors:

$$\phi(\boldsymbol{x}) = \sum_n \phi_n \psi_n(\boldsymbol{x}) \quad \text{and} \quad J(\boldsymbol{x}) = \sum_n J_n \psi_n(\boldsymbol{x}), \tag{12.4}$$

where ϕ_n and J_n are expansion coefficients specified, as usual, by

$$\phi_n = \int d^d \boldsymbol{x} \psi_n^*(\boldsymbol{x}) \phi(\boldsymbol{x}) \equiv ((\psi_n, \phi)) \quad \text{and} \quad J_n = \int d^d \boldsymbol{x} \psi_n^*(\boldsymbol{x}) J(\boldsymbol{x}) \equiv ((\psi_n, J)) \tag{12.5}$$

with asterisks denoting complex conjugates. The double parentheses on the right represent abbreviations for the indicated integrals.

As explained in conjunction with Eq. (9.7) or (9.37), $\hat{\mathcal{M}}(\boldsymbol{x}, \boldsymbol{y})$ acting on these basis vectors gives rise to the set of eigenvalue equations

$$\hat{\mathcal{M}}(\boldsymbol{x}, \boldsymbol{y}) \psi_n(\boldsymbol{x}) = \lambda_n \psi_n(\boldsymbol{x}), \quad n = 1, 2, \ldots, N, \tag{12.6}$$

where the λ_n are the eigenvalues.

In this new notation the exponent in Eq. (12.3) reads as follows:

$$\begin{aligned} E_x &= -\sum_n \sum_m ((\phi_n \psi_n, \hat{\mathcal{M}} \phi_m \psi_m)) + \sum_n \sum_m ((J_n \psi_n, \phi_m \psi_m)) \\ &= -\sum_n \sum_m ((\phi_n \psi_n, \lambda_m \phi_m \psi_m)) + \sum_n \sum_m ((J_n \psi_n, \phi_m \psi_m)) \\ &= -\sum_n (\lambda_n \phi_n^2 - J_n \phi_n). \end{aligned} \tag{12.7}$$

The latter step results from the orthonormality of the ψ_n basis vectors (the parenthesis on the bottom line is now of the usual type) that causes the integrals to vanish unless $n = m$ and reduces the double summations to a single sum over n. Notice that E_x now involves discretized basis vectors, so that (see Appendix 12.C) the measure of the integrals $\mathcal{D}\phi$ can be rewritten as a product of differentials, $\prod_{n=1}^{N} d\phi_n$.

As noted in Appendix 12.C, the original function (12.3) can then be reduced to

$$\begin{aligned} G_N[J] &= \int \cdots \int d\phi_1 \ldots d\phi_N \exp\left[-\sum_n \left(\lambda_n \phi_n^2 - J_n \phi_n\right)\right] \\ &= \prod_n \int_n d\phi_n \exp\left[-\left(\lambda_n \phi_n^2 - J_n \phi_n\right)\right], \end{aligned} \tag{12.8}$$

which will be recognized as a set of standard expanded Gaussian integrals treated in Appendix 12.A. Carrying out the integrations yields

$$\begin{aligned} G_N[J] &= \frac{\pi^{N/2}}{\left[\prod_{n=1}^{N} \lambda_n\right]^{1/2}} \exp\left[\sum_{n=1}^{N} \frac{J_n^2}{4\lambda_n}\right] \\ &= \frac{\pi^{N/2}}{[\det_N \mathcal{M}]^{1/2}} \exp\left[\sum_{n=1}^{N} \frac{J_n^2}{4\lambda_n}\right] \equiv G_N[0] \exp\left[\sum_{n=1}^{N} \frac{J_n^2}{4\lambda_n}\right], \end{aligned} \tag{12.9}$$

where we have first replaced the product under the square root by the determinant for the matrix entries of (12.6) and then separated out that part of the function that does not depend on J. This completes our digression.

12.2 PARTITION FUNCTION FOR THE GAUSSIAN MODEL

We start by rewriting the function of interest (12.1) in a form that corresponds to Eq. (12.3), so that we can use the machinery of the prior section. The rewritten Hamiltonian in Eq. (12.1) has the form

$$\hat{\mathcal{H}}_0 = \iint d^d x d^d y \phi(y) \left[\frac{1}{2}\delta(x-y)(\alpha \nabla_x^2 - u)\right] \phi(x) - \int d^d z J(z)\phi(z). \tag{12.10}$$

Notice first that here the operation $|\nabla_x \phi|^2$ of Eq. (12.1) is replaced by $-\phi(x)\nabla_x^2\phi$. This step is rendered plausible by the example worked out in Appendix 12.B. We then replace the set of single integrals over the first two terms in Eq. (12.1) by a set of double integrals that includes the delta function $\delta(x - y)$.

This allows us to split $\phi^2(\boldsymbol{x})$ into factors $\phi(\boldsymbol{x})$ and $\phi(\boldsymbol{y})$ as shown. Eq. (12.10) is now in the same format as Eq. (12.3), with the identification

$$\hat{\mathcal{M}}(\boldsymbol{x},\boldsymbol{y}) = \frac{1}{2}\delta(\boldsymbol{x}-\boldsymbol{y})(\alpha\nabla_x^2 - u). \tag{12.11}$$

By direct substitution we can verify that the eigenfunctions (omitting the unneeded normalization constants) and eigenvalues of $(\alpha\nabla_x^2 - u)\phi(\boldsymbol{x})$ are specified by

$$\psi_n(\boldsymbol{x}) = e^{i\boldsymbol{k}\cdot\boldsymbol{x}} \quad \text{and} \quad \lambda_k = -\frac{1}{2}(\alpha k^2 + u). \tag{12.12}$$

As an aside: If the eigenfunctions $\psi_n(\boldsymbol{x})$ are to "fit in" a finite volume $V = L^d$, then we must introduce periodic boundary conditions of the type $|k_{l_q}| = \frac{2\pi l_q}{L}$, l_q an integer, along every axis. This discretization should always be indicated by the subscript l_q on k. To avoid excessive notational discomfort, this subscript will be omitted; however, keep the discretization in mind.

The idea now is to use the identification of λ_k and to run the derivation from Eq. (12.3) to Eq. (12.7) backwards, so as to recast Eq. (12.3) in the form

$$\hat{\mathcal{H}}_0 = \iint d^d x d^d y \phi(\boldsymbol{y})\left[\frac{1}{2}\delta(\boldsymbol{x}-\boldsymbol{y})(\alpha k^2 + u)\right]\phi(\boldsymbol{x}) - \int d^d z J(\boldsymbol{z})\phi(\boldsymbol{z}). \tag{12.13}$$

12.3 OPERATIONS IN k SPACE

The next step involves a switch to reciprocal space, which simplifies subsequent manipulations. For this purpose, we introduce a slightly modified, commonly used version of the Fourier series for discrete $\boldsymbol{k}$ variables:

$$\begin{aligned} \phi(\boldsymbol{x}) &= \frac{1}{L^d}\sum_{k<\Lambda}\phi(\boldsymbol{k})e^{i\boldsymbol{k}\cdot\boldsymbol{x}}, \\ J(\boldsymbol{x}) &= \frac{1}{L^d}\sum_{k'<\Lambda}J(\boldsymbol{k}')e^{i\boldsymbol{k}'\cdot\boldsymbol{x}}, \end{aligned} \tag{12.14}$$

where the factor L^{-d} is introduced for later notational convenience. In what follows, we will not always indicate the restriction on k or k'.

Let us first substitute Eq. (12.14) into the second term of Eq. (12.13), designated as I_2:

$$I_2 = -\frac{1}{L^{2d}}\sum_{\boldsymbol{k}'}\sum_{\boldsymbol{k}}\int d^d z J(\boldsymbol{k}')\phi(\boldsymbol{k})\exp\left\{i(\boldsymbol{k}'+\boldsymbol{k})\cdot\boldsymbol{z}\right\}. \tag{12.15}$$

Now introduce the orthonormalization $\int d^d z \exp\{i(\boldsymbol{k}' - \boldsymbol{k}) \cdot \boldsymbol{z}\} = L^d \delta_{\boldsymbol{k}'\boldsymbol{k}}$, where $\delta_{\boldsymbol{k}'\boldsymbol{k}}$ is the Kronecker delta. Then all integrals over z except for those with $\boldsymbol{k}' - \boldsymbol{k} = 0$ are killed off. The double sum reduces to a single sum of the type

$$I_2 = -\frac{1}{L^d} \sum_{\boldsymbol{k}} \phi(\boldsymbol{k}) J(-\boldsymbol{k}). \tag{12.16}$$

The first term in Eq. (12.13) is handled similarly. An integration over $\boldsymbol{y}$ that involves $\delta(\boldsymbol{x} - \boldsymbol{y})$ eliminates that integral and changes $\phi(\boldsymbol{y})$ into $\phi(\boldsymbol{x})$. Applying Eq. (12.14), we rewrite the first term as

$$I_1 = \frac{1}{L^{2d}} \sum_{\boldsymbol{k}'} \sum_{\boldsymbol{k}} \int d^d \boldsymbol{x} \phi(\boldsymbol{k}') \phi(\boldsymbol{k}) \left[\frac{1}{2} (\alpha k^2 + u) \right] \exp\left\{ i(\boldsymbol{k}' + \boldsymbol{k}) \cdot \boldsymbol{x} \right\}. \tag{12.17}$$

Then, the integration over $\boldsymbol{x}$ yields

$$I_1 = \frac{1}{L^d} \sum_{\boldsymbol{k}} \frac{1}{2} (\alpha k^2 + u) \phi(\boldsymbol{k}) \phi(-\boldsymbol{k}), \tag{12.18}$$

so that the Hamiltonian (12.13), expressed in terms of reciprocal space variables, reads

$$\hat{\mathcal{H}}_0 = \frac{1}{L^d} \sum_{k<\Lambda} \left[\frac{1}{2} (\alpha k^2 + u) \phi(\boldsymbol{k}) \phi(-\boldsymbol{k}) - \phi(\boldsymbol{k}) J(-\boldsymbol{k}) \right]. \tag{12.19}$$

This gets us closer to the stated goal of setting up partition functions. However, we now face the problem that $\hat{\mathcal{H}}_0$ should have only real order parameter functions as its argument. This difficulty is addressed by introducing real and imaginary components in $\phi = \phi_R + i\phi_I$ and $J = J_R + iJ_I$, along with their complex conjugates $\phi^* = \phi_R - i\phi_I$ and $J^* = J_R - iJ_I$, and by requiring that $\phi(-\boldsymbol{k}) = \phi^*(\boldsymbol{k})$ and $J(-\boldsymbol{k}) = J^*(\boldsymbol{k})$. It is then easy to verify that the two distinct complex functions $\phi(\boldsymbol{k}_1)$ and $\phi(-\boldsymbol{k}_2)$ assume the form (the imaginary terms cancel out)

$$\begin{aligned} \phi(\boldsymbol{k}_1)\phi^*(-\boldsymbol{k}_2) + \phi^*(\boldsymbol{k}_1)\phi(-\boldsymbol{k}_2) &= 2\left[\phi_R(\boldsymbol{k}_1)\phi_R(-\boldsymbol{k}_2) + \phi_I(\boldsymbol{k}_1)\phi_I(-\boldsymbol{k}_2) \right] \\ &\to 2\left[\phi_R^2(\boldsymbol{k}) + \phi_I^2(\boldsymbol{k}) \right], \end{aligned} \tag{12.20}$$

where the right side obtains since $\boldsymbol{k}_1 = -\boldsymbol{k}_2 = \boldsymbol{k}$. A similar operation applies to the product $\phi(\boldsymbol{k}) J(-\boldsymbol{k})$. We have thereby reformulated $\hat{\mathcal{H}}_0$ solely in terms of real order parameters. However, only half of the $\phi(\boldsymbol{k})$ and half of the $J(\boldsymbol{k})$ are now independent variables. Thus, the sum over all $\boldsymbol{k}$ must correspondingly be reduced by half; designating this sum by $\sum^{-}$ we then reexpress Eq. (12.19) as

$$\hat{\mathcal{H}}_0 = \frac{1}{L^d} \sum_{k<\Lambda}^{-} \left\{ (\alpha k^2 + u) \left[\phi_R^2(\boldsymbol{k}) + \phi_I^2(\boldsymbol{k}) \right] - 2 \left[\phi_R(\boldsymbol{k}) J_R(\boldsymbol{k}) + \phi_I(\boldsymbol{k}) J_I(\boldsymbol{k}) \right] \right\}. \tag{12.21}$$

12.4 PARTITION FUNCTION

We now set the stage for developing the requisite partition functions. We separate out the integrations over the real and imaginary components[1] of the order parameter and switch from an exponentiated sum of terms to a product of exponents to write

$$\begin{aligned} \mathcal{Z}_0[\Lambda, J] &= \prod_{k<\Lambda}^{-} \iint d\phi_R(\boldsymbol{k}) d\phi_I(\boldsymbol{k}) e^{-\hat{\mathcal{H}}_0} \\ &= \prod_{k<\Lambda}^{-} \iint d\phi_R(\boldsymbol{k}) d\phi_I(\boldsymbol{k}) \exp\left\{ -\frac{1}{L^d} \left\{ (\alpha k^2 + u) \left[\phi_R^2(\boldsymbol{k}) + \phi_I^2(\boldsymbol{k}) \right] \right.\right. \\ &\qquad \left.\left. + 2 \left[\phi_R(\boldsymbol{k}) J_R(\boldsymbol{k}) + \phi_I(\boldsymbol{k}) J_I(\boldsymbol{k}) \right] \right\} \right\}. \end{aligned} \tag{12.22}$$

In executing first the integration over $\phi_R(\boldsymbol{k})$, which is independent of $\phi_I(\boldsymbol{k})$, this integral is seen to be in the form of the modified Gaussian integrals that we have encountered in Appendix 12.A. We find that the integral converges to

$$\mathcal{I}_R = \left[\frac{\pi L^d}{(\alpha k^2 + u)} \right]^{1/2} \times \exp\left[\frac{J_R^2(\boldsymbol{k})}{L^d (\alpha k^2 + u)} \right]. \tag{12.23}$$

A similar result is found for the integral over $\phi_I(\boldsymbol{k})$, leading to

$$\mathcal{Z}_0[\Lambda, J] = \prod_{k<\Lambda}^{-} \left[\frac{\pi L^d}{(\alpha k^2 + u)} \times \exp\left\{ \frac{1}{L^d} \frac{J_R^2(\boldsymbol{k}) + J_I^2(\boldsymbol{k})}{\alpha k^2 + u} \right\} \right]. \tag{12.24}$$

It is customary to disregard the factor πL^d because the thermodynamic functions derived from $\ln \mathcal{Z}_0$ are either taken relative to an arbitrary constant or

1. There is a semantic problem here: the functions on the right are all real but are labeled by R and by I to indicate their provenance from the complex function $\phi(\boldsymbol{k})$.

involve the differentiation of $\ln \mathcal{Z}_0$. We recast the revised expression in the form

$$\mathcal{Z}_0[\Lambda, J] = \exp\left\{-\sum_{k<\Lambda}^{-} \ln(\alpha k^2 + u)\right\} \times \exp\left\{\frac{1}{L^d}\sum_{k<\Lambda}^{-} \frac{J_R^2(\boldsymbol{k}) + J_I^2(\boldsymbol{k})}{\alpha k^2 + u}\right\}, \tag{12.25}$$

and then, in the second exponent, reverse the earlier process that had separated the complex order parameter into real and imaginary parts, so that

$$\mathcal{Z}_0[\Lambda, J] = \exp\left\{-\frac{1}{2}\sum_{k<\Lambda} \ln(\alpha k^2 + u)\right\} \times \exp\left\{\frac{1}{2L^d}\sum_{k<\Lambda} \frac{J(\boldsymbol{k})J(-\boldsymbol{k})}{\alpha k^2 + u}\right\}, \tag{12.26}$$

where the unrestricted summations have now been reintroduced, along with the requisite factor $\frac{1}{2}$. In the limit of replacing discrete by continuum variables the summations may be replaced by integrations, yielding

$$\mathcal{Z}_0[\Lambda, J] = \exp\left\{-\frac{V}{2}\int_{0\le k\le\Lambda} d^d\boldsymbol{k}\, \ln(\alpha k^2 + u)\right\} \times \exp\left\{\frac{1}{2}\int_{0\le k\le\Lambda} d^d\boldsymbol{k}\frac{J(\boldsymbol{k})J(-\boldsymbol{k})}{\alpha k^2 + u}\right\}. \tag{12.27}$$

Notice that we have now split the partition function into a part that depends on J, separate from an independent multiplier. We have also abandoned the $\boldsymbol{k}$ quantization.

This gets us to our ultimate destination: We can now specify the Helmholtz free energy through the standard relation $F_H[\Lambda, J] = -\beta^{-1}\ln \mathcal{Z}_0[\Lambda, J]$. Thus,

$$F_H[\Lambda, J] = \beta^{-1}\left\{\frac{V}{2}\int_{0\le k\le\Lambda} d^d\boldsymbol{k}\, \ln(\alpha k^2 + u) - \frac{1}{2}\int_{0\le k\le\Lambda} d^d\boldsymbol{k}\frac{J(\boldsymbol{k})J(-\boldsymbol{k})}{\alpha k^2 + u}\right\}, \tag{12.28}$$

which is in the format of Eq. (12.9).

12.5 CORRELATION FUNCTIONS

We are also ready for the specification of expectation values and correlation functions. Return to Eq. (12.27) in the form

$$\mathcal{Z}_0[\Lambda, J] = \mathcal{Z}_0[\Lambda, 0]\exp\left\{\frac{1}{2}\int_{0\le k\le\Lambda} d^d\boldsymbol{k}\frac{J(\boldsymbol{k})J(-\boldsymbol{k})}{\alpha k^2 + u}\right\} \tag{12.29}$$

and introduce the Fourier relation (corresponding to the inverse to Eq. (12.14)) that specifies $J(\boldsymbol{k})$ in terms of $J(\boldsymbol{x})$, taking note of the L^{-d} factor implicit in

$\mathcal{Z}_0[\Lambda, 0]$. Set

$$J(\boldsymbol{k}) = (2\pi)^{-d} \int d^d\boldsymbol{x}\, J(\boldsymbol{x}) e^{-i\boldsymbol{k}\cdot\boldsymbol{x}}, \tag{12.30}$$

and introduce this relation in (12.29). This gets us back to the direct space in specifying $\mathcal{Z}_0$. The integral in that expression now assumes the form

$$\begin{aligned} I_z &= \int_{0\le k\le\Lambda} \frac{d^d\boldsymbol{k}}{(2\pi)^{2d}(\alpha k^2+u)} \int d^d\boldsymbol{y}\, J(\boldsymbol{y}) e^{-i\boldsymbol{k}\cdot\boldsymbol{y}} \int d^d\boldsymbol{x}\, J(\boldsymbol{x}) e^{i\boldsymbol{k}\cdot\boldsymbol{x}} \\ &= \iint d^d\boldsymbol{x} d^d\boldsymbol{y}\, J(\boldsymbol{x}) \left[\int_{0\le k\le\Lambda} \frac{d^d\boldsymbol{k}}{(2\pi)^{2d}} \frac{e^{i\boldsymbol{k}\cdot[\boldsymbol{x}-\boldsymbol{y}]}}{\alpha k^2+u} \right] J(\boldsymbol{y}) \\ &\equiv \iint d^d\boldsymbol{x} d^d\boldsymbol{y}\, J(\boldsymbol{x}) \Delta(\boldsymbol{x}-\boldsymbol{y}) J(\boldsymbol{y}), \end{aligned} \tag{12.31}$$

where, using conventional notation, we have introduced a new integral, namely

$$\Delta(\boldsymbol{x}-\boldsymbol{y}) \equiv \int_{0\le k\le\Lambda} \frac{d^d\boldsymbol{k}}{(2\pi)^{2d}} \frac{e^{i\boldsymbol{k}\cdot[\boldsymbol{x}-\boldsymbol{y}]}}{\alpha k^2+u}. \tag{12.32}$$

This quantity has an important function in our further development, as we now discover and will also use in the future.

In what follows, we need the functional differentiation of $J(\boldsymbol{x})$, alluded to in Appendix 12.D, that involves the symbolic relation $\frac{\partial J(\boldsymbol{x})}{\partial J(\boldsymbol{y})} = \delta(\boldsymbol{x}-\boldsymbol{y})$, which introduces the conventional Dirac delta function. Then, according to the discussion of Chapter 3, concerning connected correlations, we can construct the expectation value for the connected correlation $G_c^{(1)}(\boldsymbol{x}) \equiv \langle\phi(\boldsymbol{x})\rangle$ as

$$\begin{aligned} G_c^{(1)}(\boldsymbol{x}) &= \frac{\partial}{\partial J(\boldsymbol{x})} \ln \mathcal{Z}_0[\Lambda, J] = 2 \times \frac{1}{2} \int d^d\boldsymbol{y}\, \Delta(\boldsymbol{x}-\boldsymbol{y}) J(\boldsymbol{y}) \\ &= \int d^d\boldsymbol{y}\, \Delta(\boldsymbol{x}-\boldsymbol{y}) J(\boldsymbol{y}) = \langle\phi(\boldsymbol{x})\rangle. \end{aligned} \tag{12.33}$$

This thermal average vanishes if $J = 0$. The factor 2 arises from differentiations of both $J(\boldsymbol{x})$ and $J(\boldsymbol{y})$, with $\boldsymbol{x}$ and $\boldsymbol{y}$ interchanged.

In a similar manner, the connected correlation function is seen to be

$$\begin{aligned} G_c^{(2)}(\boldsymbol{x}_1, \boldsymbol{x}_2) &= \frac{\partial}{\partial J(\boldsymbol{x}_1)} \frac{\partial}{\partial J(\boldsymbol{x}_2)} \frac{1}{2} \iint d^d\boldsymbol{x} d^d\boldsymbol{y}\, J(\boldsymbol{x}) \Delta(\boldsymbol{x}-\boldsymbol{y}) J(\boldsymbol{y}) \\ &= \frac{1}{2} \iint d^d\boldsymbol{x} d^d\boldsymbol{y} \frac{\partial J(\boldsymbol{x})}{\partial J(\boldsymbol{x}_1)} \Delta(\boldsymbol{x}-\boldsymbol{y}) \frac{\partial J(\boldsymbol{y})}{\partial J(\boldsymbol{x}_2)} \\ &\quad + \frac{1}{2} \iint d^d\boldsymbol{x} d^d\boldsymbol{y} \frac{\partial J(\boldsymbol{x})}{\partial J(\boldsymbol{x}_2)} \Delta(\boldsymbol{x}-\boldsymbol{y}) \frac{\partial J(\boldsymbol{y})}{\partial J(\boldsymbol{x}_1)}, \end{aligned} \tag{12.34}$$

so that, interchanging integration variables $\boldsymbol{x}$ and $\boldsymbol{y}$. we find that

$$G_c^{(2)}(\boldsymbol{x}_1, \boldsymbol{x}_2) = \iint d^d\boldsymbol{x} d^d\boldsymbol{y} \delta(\boldsymbol{x}-\boldsymbol{x}_1)\Delta(\boldsymbol{x}-\boldsymbol{y})\delta(\boldsymbol{y}-\boldsymbol{x}_2) = \Delta(\boldsymbol{x}_1-\boldsymbol{x}_2). \tag{12.35}$$

We have now specified the connected correlation function. Higher-order connected correlations vanish.

12.6 COMMENT

As a commentary on the above operations, we take up convergence problems by introducing Eq. (12.32) into (12.28), thus reintroducing the direct space into the second term:

$$F_{\hat{\mathcal{H}}}[\Lambda, J] = \beta^{-1}\left\{\frac{V}{2}\int_{0\le k\le\Lambda} d^d\boldsymbol{k}\ln(\alpha k^2+u) - \frac{1}{2}\iint d^d\boldsymbol{x} d^d\boldsymbol{y} J(\boldsymbol{x})\Delta(\boldsymbol{x}-\boldsymbol{y})J(\boldsymbol{y})\right\}. \tag{12.36}$$

In general, as $\Lambda \to \infty$, the integrals diverge, which is avoided here by rendering Λ finite. However, the first term in (12.36) depends strongly on Λ. Increasing Λ drags more degrees of freedom into the model that contribute to the free energy. Large Λ values in turn imply only small ranges of the distance Λ^{-1} that can be retained, that are close to the microscopic domain, precisely what we were trying to avoid in the first place. The second term does not depend strongly on Λ so long as $J(\boldsymbol{x})$ remains smooth in the range greater than Λ^{-1}. Since it is $J(\boldsymbol{x})$ that is of interest in dealing with critical phenomena, the problems associated with the first term may not be relevant. Also, we are ultimately not interested so much in the microscopic parameters α, u, λ , $\Lambda, \ldots$ as in their relation to macroscopic observables such as temperature, correlation length, magnetic properties, etc., where the microscopic irregularities are automatically smoothed out.

Other difficulties lurk in the previous formulations: Clearly, we must have $u > 0$ to prevent the unpleasantness of a vanishing denominator, or of a negative logarithmic argument. Also, aside from the numerator $J(\boldsymbol{k})J(-\boldsymbol{k})$, when $u \ll \alpha k^2$, the integral $\sim \int_{0\le k\le\Lambda} dk \frac{k^{d-1}}{\alpha k^2+u}$ is proportional to Λ^{d-2}, which diverges for $d \ge 3$ as $\Lambda \to \infty$. This is known as the *ultraviolet divergence*. On the other hand, if $d \le 2$ and $u = 0$, then the integral diverges as $k \to 0$, which is known as the *infrared divergence*. This is a prime example of the significant role played by the spatial dimension.

12.7 RENORMALIZATION OF THE HAMILTONIAN

We now build on the preceding discussion to examine conditions under which a fixed point might be reached. For simplicity, we start with a limited Hamiltonian in the absence of a magnetic field, as specified by Eq. (12.19), but generalized to the continuum, that is, with the summation replaced by an integration. We write

$$\hat{\mathcal{H}}_0 = \int d^d\boldsymbol{k}\frac{1}{2}(u+\alpha k^2)\phi(\boldsymbol{k})\phi(-\boldsymbol{k}) = \int d^d\boldsymbol{k}\frac{1}{2}(u+\alpha k^2)\,|\phi(\boldsymbol{k})|^2. \quad (12.37)$$

The problem now is to find out precisely how the Hamiltonian is altered when a renormalization step is carried out. This matter is discussed at various levels of sophistication in all standard texts; here we pursue a heuristic argument that will be properly firmed up in Chapter 13. If we are to compare the original with the renormalized Hamiltonian, then we must allow the integration step to encompass the range $0 \le k \le \Lambda/b$, which corresponds to the block representation of the lattice. Thus, we write

$$\hat{\mathcal{H}}_0 = \int_{k\le\Lambda/b} d^d\boldsymbol{k}\frac{1}{2}\left(u+\alpha k^2\right)\phi(\boldsymbol{k})\phi(-\boldsymbol{k}) = \int_{k\le\Lambda/b} d^d\boldsymbol{k}\frac{1}{2}\left(u+\alpha k^2\right)|\phi(\boldsymbol{k})|^2. \quad (12.38)$$

We now adjust the parameters such that the renormalized Hamiltonian exhibits the same functional structure as the original. Thus, starting with Eq. (12.37), we introduce the scaling relations established in Eqs. (11.12) and (11.13):

$$k = \frac{k'}{b}; \quad \phi(\boldsymbol{k}) = b^{d-d_\phi}\phi'(\boldsymbol{k}'). \quad (12.39)$$

Substituting Eq. (12.38) into Eq. (12.37), so as to rescale the Hamiltonian, we find that

$$\hat{\mathcal{H}}_0' = \int_{k'<\Lambda} d^d\boldsymbol{k}' b^{-d}\frac{1}{2}\left(u+\frac{\alpha k'^2}{b^2}\right)b^{2(d-d_\phi)}|\phi'(\boldsymbol{k}')|^2, \quad (12.40)$$

which may be compared with Eq. (12.37) to establish the correspondence

$$u' = b^{(d-2d_\phi)}u \quad \text{and} \quad \alpha' = b^{(d-2d_\phi-2)}\alpha. \quad (12.41)$$

12.8 ESTABLISHING THE REQUIREMENTS OF THE FIXED POINT

We can now study the conditions required to reach a fixed point characterized by the requirements $u = u'$ and $\alpha = \alpha'$. Eqs. (12.40) allow for two possibilities with $b > 1$:

We may set $d - 2d_\phi = 0$; then $u = u'$ and $\alpha' = b^{-2}\alpha$. The first relation shows that u remains unchanged no matter what the RGT operation is, and thus provides no information; any value is consistent with the fixed point. Under repeated rescalings with $b > 1$, the latter relation shows that α approaches zero, corresponding to setting $y_1 = -2$ in Eq. (9.12). This renders α an irrelevant parameter; we have reached a critical point. However, as follows from inspection of Eq. (11.5), $\alpha = 0$ implies the absence of any nearest-neighbor interactions. At this critical point, we thus deal with a lattice of noninteracting sites, corresponding to the limit of infinite temperatures, with an anomalous dimension $d_\phi = d/2$.

We may set $d - 2d_\phi - 2 = 0$; then $\alpha' = \alpha$ and $u' = b^2 u$. Here the value of α is immaterial, whereas u grows under repeated rescaling; with $y = 2 > 0$ in Eq. (9.12), u becomes a relevant variable. It also follows that $\nu = 1/2$. To reach the critical point, we must force the condition $u = 0$. Since, according to Eq. (11.5), u controls the phase transition, we conclude that we, in fact, do not encounter a critical point at finite temperatures. Also, the anomalous dimension now reads $d_\phi^0 = d/2 - 1$, which may be compared to the general result, Eq. (11.14), $d_\phi = d/2 - 1 + \frac{\eta}{2}$. Thus, the present case corresponds to setting $\eta = 0$. The value d_ϕ^0 is termed the *normal or canonical anomalous scaling dimension*.

12.9 CONCLUSION

In reviewing the preceding, we must face the unpleasant fact that the Gaussian model does not lead to a satisfactory description of critical phenomena at finite temperatures. In what follows, we must therefore learn how to proceed beyond the current approximation.

APPENDIX 12.A HUBBARD–STRATONOVICH TRANSFORMATION

Consider a slightly altered version of Eq. (12.8), which is made up of a product of N integrals, each of which has the general Gaussian form

$$I_n = \int_{-\infty}^{\infty} d\phi_n \exp\left[-\frac{\lambda_n\phi_n^2 - \phi_n J_n}{L}\right]. \tag{12.A.1}$$

To handle this integral, we start with the Poisson integral $\int_{-\infty}^{\infty} dy e^{-y^2} = \sqrt{\pi}$ and introduce the shift of variable $y = \frac{x}{\sqrt{2}} - a$, hence $dy = \frac{1}{\sqrt{2}}dx$, from which we obtain the important identity

$$e^{a^2} = (2\pi)^{-1/2}\int_{-\infty}^{\infty} \exp\left[-\frac{x^2}{2} + \sqrt{2}ax\right] dx. \tag{12.A.2}$$

This is an example of the so-called *Hubbard–Stratonovich transformation* that plays an important role in different branches of theoretical physics. It replaces the exponentiated square of a parameter a by an exponentiated first power of that parameter under an integral sign. The above relation can be converted to the particular expression (12.A.1) for I_n by setting $x_n = \sqrt{2\lambda_n/L}\phi_n, dx_n = \sqrt{2\lambda_n/L}d\phi_n, a_n = \frac{J_n}{2\sqrt{\lambda_n L}}$, which leads to the relation

$$\exp(a_n^2) = \sqrt{\frac{\lambda_n}{\pi L}} I_n, \qquad (12.A.3)$$

where I_n is specified by Eq. (12.A.1). Then

$$I_n\left[J, \lambda_n\right] = \sqrt{\frac{\pi L}{\lambda_n}} \exp\left[\frac{J_n^2}{4\lambda_n L}\right]. \qquad (12.A.4)$$

Since Eq. (12.9) requires a concatenation of N such products, we rewrite the above as

$$\begin{aligned} I_N\left[J, \lambda_n\right] &= \frac{(\pi L)^{\frac{N}{2}}}{\left\{\prod_{n=1}^{N} \lambda_n\right\}^{\frac{1}{2}}} \prod_{n=1}^{N} \exp\left[\frac{J_n^2}{4\lambda_n L}\right] \\ &= \frac{(\pi L)^{\frac{N}{2}}}{\left\{\prod_{n=1}^{N} \lambda_n\right\}^{\frac{1}{2}}} \exp\left\{\sum_{n=1}^{N}\left[\frac{J_n^2}{4\lambda_n L}\right]\right\}, \end{aligned} \qquad (12.A.5)$$

which is the desired relation.

APPENDIX 12.B RELATION $|\nabla_x \phi|^2 = -\phi\nabla^2\phi$

We investigate the replacement of $|\nabla_x\phi|^2$ by $-\phi\nabla_x^2\phi$ under an integral by resorting to a one-dimensional analogy. Consider the following integral and execute an integration by parts:

$$\int_{-\infty}^{\infty} dx \left(f\frac{\partial^2 f}{\partial x^2}\right) = \left[f\frac{\partial f}{\partial x}\right]_{-\infty}^{\infty} - \int_{-\infty}^{\infty} dx \left(\frac{\partial f}{\partial x}\right)^2. \qquad (12.B.1)$$

If either f or $\frac{\partial f}{\partial x}$ vanishes at $x = -\infty$ and $x = \infty$, then the first term on the right drops out and Eq. (12.B.1) reduces to

$$\int_{-\infty}^{\infty} dx \left(f\frac{\partial^2 f}{\partial x^2}\right) = - \int_{-\infty}^{\infty} dx \left(\frac{\partial f}{\partial x}\right)^2. \qquad (12.B.2)$$

The extension to higher dimensions is left as an exercise.

APPENDIX 12.C COMMENT ON FUNCTIONAL INTEGRATION

The term "functional integration" as used in the physics community is best explained by analogy to an ordinary integral of the type $\int g(x,y)dx\,dy$, where we sum $g(x,y)$ as we raster x and y over sets of cells of size $dxdy$. A functional integral involves an operation in which a functional (a function of a function) changes as its argument (a function) rasters over the set of *all* relevant (or admissible) functions, each multiplied by the "volume" element it occupies in the function space. In carrying this out in the manner shown below, we should deal with an infinite set of real numbers that must be integrated over, requiring an infinite number of integrals. This raises obvious problems.

So, how do we do this? By way of an example pertinent to the present chapter, consider the functional

$$\mathcal{F}\left[g\right] = \exp\left[-c\int_{-L/2}^{L/2} dx g^2(x)\right], \tag{12.C.1}$$

involving an exponentiation of integrals of functions $g(x)$ in the range $-\frac{L}{2} \leq x \leq \frac{L}{2}$. The corresponding *functional integral* has the form

$$\mathcal{L} = \int \mathcal{D}g \exp\left[-c\int_{-L/2}^{L/2} dx g^2(x)\right], \tag{12.C.2}$$

where $\int \mathcal{D}g$ means that the integration is to include all admissible functions $g(x)$ in the indicated interval.

To actually execute this operation, we break up the interval $\left[-L/2, L/2\right]$ into $N-1$ equal segments and discretize every function in the set $g(x)$ as follows:

$$f\left\{g_i\right\} = \exp\left[-\frac{cL}{N}\sum_{i=1}^{N} g_i^2\right], \tag{12.C.3}$$

where $g_i \equiv g((\frac{i}{N} - \frac{1}{2})L)$, and the summation covers the indicated x range. We then add up the contributions from all the functions included in the $g(x)$ set, writing

$$\mathcal{L}_N = \int \ldots \int dg_1 \ldots dg_N \exp\left[-\frac{cL}{N}\sum_{i=1}^{N} g_i^2\right]. \tag{12.C.4}$$

This now has the form of a set of ordinary integrals. As usual, we convert the exponentiated sum into a product of exponentials, so that

$$\mathcal{L}_N = \prod_{i=1}^{N} \int dg_i \exp\left[-\left(\frac{cL}{N}\right) g_i^2\right] = \left(\frac{\pi N}{cL}\right)^{\frac{N}{2}}, \tag{12.C.5}$$

where the standard expression for the Gaussian integral was introduced on the right. As expected in any integration, the final result makes no reference to g_i.

We should now go to the limit as $N \to \infty$, so that Eq. (12.C.5) converges to Eq. (12.C.1). Here we are faced with the problem of a diverging result. However, in problems of interest in condensed matter physics, where atoms are the fundamental building blocks, we ordinarily do not encounter true continuities. Thus, in our case, the divergence problem can be cured by adopting, as discussed earlier, restrictions on the set of admissible functions: eliminating those that are not "well behaved" or those involving the range of atomic and subatomic dimensions. In fact, in our case, either when dealing with a finite-dimensional lattice of spins, or via the use of a finite number of basis functions in Eq. (12.4), the discretization is already done for us.

APPENDIX 12.D MEANING OF FUNCTIONAL (VARIATIONAL) DERIVATIVES

Here we provide a cursory glance at functional differentiations for one-dimensional systems. Notice how we relate, via the discretization process, the functional to the ordinary differentiation:

Consider a functional I of a set of functions $\phi(x)$ that is formed in the continuum limit of N variables ϕ_i. For the particular case of only one dimension, the basic definition of the functional differentiation for $\frac{\delta I}{\delta\phi(x)}$ reads

$$\frac{\delta \mathcal{I}}{\delta\phi(x)} = \lim_{c\to 0} \frac{1}{c}\frac{\partial \mathcal{I}}{\partial \phi_i}. \tag{12.D.1}$$

We can now proceed to check out some examples.

Consider the case where $\mathcal{I} = \int f(z)\phi^p(z)dz$, which is to be taken as $\mathcal{I} = \lim_{c\to 0} c\sum_l f_l\phi_l^r$. Then $\partial\mathcal{I}/\partial\phi_s = crf_s\phi_s^{r-1}$, whence

$$\frac{\delta \mathcal{I}}{\delta\phi(z)} = rf\phi^{r-1}(z). \tag{12.D.2}$$

This feature distinguishes the functional from the ordinary derivative. Notice the elimination of the integral sign. The special case $\mathcal{I} = \int f[\phi(z)]dz$ to yield

$\frac{\partial \mathcal{I}}{\partial \phi(y)} = \frac{d}{dy} f[\phi(y)]$ can be easily established. It is then easy to check the following results: Based on the fact that $\partial \phi_l / \partial \phi_m = \delta_{lm}$, we have

$$\frac{\delta}{\delta \phi(\boldsymbol{r})} \phi(\boldsymbol{r}') = \delta(\boldsymbol{r} - \boldsymbol{r}') \equiv \frac{\delta}{\delta \phi(\boldsymbol{r})} \int d^d r' \, \delta(\boldsymbol{r} - \boldsymbol{r}') \phi(\boldsymbol{r}'). \tag{12.D.3}$$

Also,

$$\frac{\delta}{\delta \phi(\boldsymbol{r})} \int d^d \boldsymbol{r}' \phi(\boldsymbol{r}') = 1 \equiv \frac{\delta}{\delta \phi(\boldsymbol{r})} \int d^3 r' \int d^3 r \, \phi(\boldsymbol{r}) \delta(\boldsymbol{r} - \boldsymbol{r}'). \tag{12.D.4}$$

Chapter 13

The Ginzburg–Landau–Wilson Formalism: Beyond the Gaussian Approximation

At this point, we take the next step by going beyond the zero-order methodology of the Gaussian model.

13.1 GENERALITIES

We follow the exposition by W.D. McComb[1] by considering the model Hamiltonian

$$\hat{\mathcal{H}} = \int d^d\boldsymbol{x}\left[\frac{1}{2}(\nabla\varphi)^2 + \frac{u}{2}\varphi^2 + \frac{\lambda}{4!}\varphi^4\right] \equiv \hat{\mathcal{H}}_0 + \hat{\mathcal{H}}_1. \quad (13.1)$$

As a reminder: We have set the nearest-neighbor interaction parameter $\alpha = 1$; $u > 0$ depends on the temperature T and governs the phase transition; λ is a coupling parameter that introduces the fourth power of the position, dependent order parameter φ. Unfortunately, we can no longer factorize the this operator in the reciprocal space, as we did earlier. As a new strategy, we break the order parameter into two pieces as follows:

$$\varphi(\boldsymbol{x}) = \varphi^+(\boldsymbol{x}) + \varphi^-(\boldsymbol{x}), \quad (13.2)$$

in which we now introduce the truncated inverse Fourier transforms

$$\begin{aligned} \varphi^+(\boldsymbol{x}) &= \int_{\frac{\Lambda}{b}\le k\le\Lambda} \frac{d^d\boldsymbol{k}}{(2\pi)^d}\left\{\exp\left[i\boldsymbol{k}\cdot\boldsymbol{x}\right]\right\}\varphi(\boldsymbol{k}) \\ \text{and } \varphi^-(\boldsymbol{x}) &= \int_{0\le k\le\Lambda/b} \frac{d^d\boldsymbol{k}}{(2\pi)^d}\left\{\exp\left[i\boldsymbol{k}\cdot\boldsymbol{x}\right]\right\}\varphi(\boldsymbol{k}). \end{aligned} \quad (13.3)$$

As is evident, the two integrals involve regions in $\boldsymbol{k}$ space[2] that correspond respectively to distances at or below the atomic domain and to distances in

1. W.D. McComb, *Renormalization Methods* (Clarendon Press, Oxford, 2004), Sec. 9.10.
2. Keep in mind that, appearances to the contrary, we are dealing with d-dimensional multiple integrals; each of the d components of the wavevector $\boldsymbol{k}$ is subject to the indicated limits.

A Primer to the Theory of Critical Phenomena. http://dx.doi.org/10.1016/B978-0-12-804685-2.00013-9

the macroscopic domain in the direct space. As explained below, rather than discarding the terms in $\varphi^+(\boldsymbol{x})$ outright, as was done in Chapter 12, we will gradually eliminate the terms involving $\varphi^+(\boldsymbol{x})$. In what follows, we consider $\hat{\mathcal{H}}_0$, defined by the first two terms in Eq. (13.1), to constitute the basic portion of the Hamiltonian and to regard the $\hat{\mathcal{H}}_1$ component, involving $\frac{\lambda\varphi^4}{4!}$, as a perturbation on $\hat{\mathcal{H}}_0$; initially, we confine ourselves to the first-order correction.

13.2 THE RGT EXECUTION

We adopt the usual methodology of gradually eliminating microscopic variables. To facilitate the flow of basic information, we start with the discussion in Appendix 13.A, using Eq. (13.A.2) as our point of departure, and execute an expansion that linearizes the perturbation term:

$$\begin{aligned} e^{-\hat{\mathcal{H}}'[\varphi^-]} &= \int \mathcal{D}\varphi^+ e^{-\hat{\mathcal{H}}[\varphi^- + \varphi^+]} \\ &\approx \int \mathcal{D}\varphi^+ e^{-\hat{\mathcal{H}}_0[\varphi^- + \varphi^+]}(1 - \hat{\mathcal{H}}_1[\varphi^- + \varphi^+]) + \cdots . \end{aligned} \tag{13.4}$$

The fact that the left-hand side involves solely $\varphi^-(\boldsymbol{x})$ as the independent variable will ultimately be verified.

The objective now is to eliminate the microscopic components associated with φ^+ before the rescaling, so that we end up dealing with φ^- alone. $\hat{\mathcal{H}}_0\varphi^-$ is separable, so that the exponential factor $e^{-\hat{\mathcal{H}}_0[\varphi^-]}$ can be moved outside the integral. Then to first order,

$$e^{-\hat{\mathcal{H}}'[\varphi^-]} = e^{-\hat{\mathcal{H}}_0[\varphi^-]} \int \mathcal{D}\varphi^+ e^{-\hat{\mathcal{H}}_0[\varphi^+]} \left(1 - \hat{\mathcal{H}}_1\left[\varphi^- + \varphi^+\right]\right), \tag{13.5}$$

which we rewrite as

$$\begin{aligned} e^{-\hat{\mathcal{H}}'[\varphi^-]} &= e^{-\hat{\mathcal{H}}_0[\varphi^-]} \left\{ \frac{\int \mathcal{D}\varphi^+ e^{-\hat{\mathcal{H}}_0[\varphi^+]} - \int \mathcal{D}\varphi^+ e^{-\hat{\mathcal{H}}_0[\varphi^+]} \hat{\mathcal{H}}_1\left[\varphi^- + \varphi^+\right]}{\int \mathcal{D}\varphi^+ e^{-\hat{\mathcal{H}}_0[\varphi^+]}} \right\} \\ &\quad \times \int \mathcal{D}\varphi^+ e^{-\hat{\mathcal{H}}_0[\varphi^+]} = \\ &= e^{-\hat{\mathcal{H}}_0[\varphi^-]} \left[1 - \frac{\int \mathcal{D}\varphi^+ e^{-\hat{\mathcal{H}}_0[\varphi^+]} (\hat{\mathcal{H}}_1\left[\varphi^- + \varphi^+\right])}{\int \mathcal{D}\varphi^+ e^{-\hat{\mathcal{H}}_0[\varphi^+]}} \right] \int \mathcal{D}\varphi^+ e^{-\hat{\mathcal{H}}_0[\varphi^+]}. \end{aligned} \tag{13.6}$$

The constant integral to the right of the square bracket may be ignored since via the $\ln \mathcal{Z}$ term it contributes only an additive constant to the free energy, which itself is known only to within an arbitrary constant. We next recast the relation inside the square brackets back to its original exponential form, so that

$$\exp(-\hat{\mathcal{H}}'[\varphi^-]) = \exp(-\hat{\mathcal{H}}[\varphi^-]) \times \exp\left\{-\frac{\int D\varphi^+ e^{-\hat{\mathcal{H}}_0[\varphi^+]}(\hat{\mathcal{H}}_1[\varphi^- + \varphi^+])}{\int D\varphi^+ e^{-\hat{\mathcal{H}}_0[\varphi^+]}}\right\}. \tag{13.7}$$

Then, the Hamiltonian before and after rescaling is related as follows:

$$\begin{aligned}\hat{\mathcal{H}}'[\varphi^-] &= \hat{\mathcal{H}}_0[\varphi^-] + \frac{\int D\varphi^+ e^{-\hat{\mathcal{H}}_0[\varphi^+]}(\hat{\mathcal{H}}_1[\varphi^- + \varphi^+])}{\int D\varphi^+ e^{-\hat{\mathcal{H}}_0[\varphi^+]}} \\ &\equiv \hat{\mathcal{H}}_0[\varphi^-] + \left\langle \hat{\mathcal{H}}_1[\varphi^- + \varphi^+] \right\rangle_0^+ .\end{aligned} \tag{13.8}$$

Here the subscript and superscript on the right remind us that the Gaussian $\langle\rangle$ averaging involves only the zero-order ground state and an averaging solely over the high wave-number states.

Next, for averaging purposes, we identify $\hat{\mathcal{H}}_1[\varphi^- + \varphi^+]$ with the φ^4 term in Eq. (13.1) and execute the fourth-power expansion as

$$\langle(\varphi^- + \varphi^+)^4\rangle_0^+ = (\varphi^-)^4 + 6(\varphi^-)^2\langle(\varphi^+)^2\rangle_0^+ + \langle(\varphi^+)^4\rangle_0^+, \tag{13.9}$$

where[3] $\langle\varphi^-\rangle_0^+ = \varphi^-$ and $\langle\varphi^+\rangle_0^+ = \langle(\varphi^+)^3\rangle_0^+ = 0$.

The third term in Eq. (13.9) does not involve φ^- and may therefore be discarded. In the central term, we make the identification

$$\langle(\varphi^+)^2\rangle_0^+ \equiv G_0^+(0), \tag{13.10}$$

where $G_0^+(0)$ is known from the zero-order theory.

We also need to examine the multiplier of $u/2$ in Eq. (13.1), namely $\frac{u}{2}(\varphi^- + \varphi^+)^2 = \frac{u(\varphi^-)^2}{2}$. Here the cross term in the expansion vanishes by reason of symmetry, and $(\varphi^+)^2$ generates a discardable constant term. Specifically, based on Eqs. (13.1), (13.9), and (13.10), we find that

$$\hat{\mathcal{H}}'[\varphi^-] = \int d^d\boldsymbol{x}\left[\frac{1}{2}(\nabla\varphi^-)^2 + \frac{1}{2}\left(u + \frac{\lambda}{2}G_0^+(0)\right)(\varphi^-)^2 + \frac{\lambda}{4!}(\varphi^-)^4\right]. \tag{13.11}$$

Observe that $G_0^+(0)$, associated via Eq. (13.9) with the multiplier $\frac{\lambda}{4!}$ of Eq. (13.1), occurs with a factor of $1/4$ (without the ! sign).

We have now managed to banish φ^+ from the Hamiltonian in Eq. (13.1).

3. The first term in Eq. (13.9) results from the fact that we average the low k frequency φ^- over the high k integral; this factor may be moved outside the Gaussian average $\langle\rangle$, which thereby equals unity. The odd-order moments of the high frequency φ^+ averages vanish by reason of symmetry.

13.3 RESCALING EFFECTS

We can complete the rescaling by refurbishing $\hat{\mathcal{H}}'[\varphi^-]$ with altered coupling coefficients, so that it can be directly compared with $\hat{\mathcal{H}}_0[\varphi^-]$. This is achieved by replacing the unprimed by primed variables, in accordance with our earlier prescriptions: with $\varphi'(\boldsymbol{x}') = b^{d_\varphi}\varphi(\boldsymbol{x})$, $\int d^d\boldsymbol{x}' = b^{-d}\int d^d\boldsymbol{x}$, and $\nabla_x = b^{-1}\nabla_{x'}$, we rewrite Eq. (13.11) in the form

$$\hat{\mathcal{H}}'[\varphi'] = \int d^d\boldsymbol{x}'\left[\frac{1}{2}b^{d-2-2d_\varphi}(\nabla'\varphi')^2 + \frac{1}{2}b^{d-2d_\varphi}\left(u + \frac{\lambda}{2}G_0^+(0)\right)(\varphi')^2 + \frac{\lambda}{4!}b^{d-4d_\varphi}(\varphi')^4\right]. \tag{13.12}$$

We want the gradient term to have the same form as before renormalization; recall that in Chapter 12 we set $\alpha = 1$. Hence, we require

$$b^{d-2-2d_\varphi} = 1, \quad \text{or} \quad d-2-2d_\varphi = 0, \quad \text{or} \quad 2d_\varphi = d-2. \tag{13.13}$$

For further progress, we must consider several cases.

13.4 CASE $d > 4$

We complete the process by rendering the renormalized Hamiltonian invariant under rescaling; that is, along with Eq. (13.13), we require that the multipliers of $(\varphi')^2$ and $(\varphi')^4$ assume the form

$$u' = b^2\left(u + \frac{\lambda}{2}G_0^+(0)\right) \quad \text{and} \quad \lambda' = b^{4-d}\lambda \equiv b^{\epsilon}\lambda, \tag{13.14}$$

where the symbol $\epsilon \equiv 4-d$ is in almost universal use because the dimension $d = 4$ plays a central role in our further development. In the present case, $\epsilon < 0$; repeated scaling decreases λ. As $\lambda \to 0$, $u' \to b^2u^*$; with $b > 1$, we must force $u \to 0$ to reach the fixed point.

To determine $G_0^+(0)$, we revert to Eqs. (12.32) and (12.35), where we had introduced and specified the function $\Delta(\boldsymbol{x}-\boldsymbol{y})$ in terms of the inverse Fourier transform

$$G_0(\boldsymbol{x}-\boldsymbol{y}) = \int_{0\le k\le\Lambda}\frac{d^d\boldsymbol{k}}{(2\pi)^{2d}}\frac{e^{i\boldsymbol{k}\cdot(\boldsymbol{x}-\boldsymbol{y})}}{\alpha k^2+u}, \tag{13.15}$$

which in our case specializes to ($\alpha = 1$)

$$G_0(0) = \int_{0\le k\le\Lambda}\frac{d^d\boldsymbol{k}}{(2\pi)^{2d}}\frac{1}{k^2+u} \tag{13.16}$$

and to the more restricted form

$$G_0^+(0) = \int_{\frac{\Lambda}{b} \le k \le \Lambda} \frac{d^d \mathbf{k}}{(2\pi)^{2d}} \frac{1}{k^2 + u}. \tag{13.17}$$

We must now deal with this integral. In the limit as $u \to 0$ (see above), the integrand is hyperspherically symmetric. Then $d^d \mathbf{k} = \frac{\mathfrak{S}_d}{(2\pi)^d} k^{d-1} dk$, where $\mathfrak{S}_d$ is the surface area of the d-dimensional hypersphere. The leading term with $u \to 0$ is obtained by straightforward integration:

$$G_0^+(0) = \frac{\mathfrak{S}_d \Lambda^{d-2}}{(2\pi)^{3d}(d-2)}(1 - b^{2-d}) \equiv 2C(1 - b^{2-d}). \tag{13.18}$$

Now introduce this relation into Eq. (13.14). This leads to the transformations

$$u' = b^2 u + C(b^2 - b^\epsilon)\lambda, \tag{13.19}$$

$$\lambda' = b^\epsilon \lambda, \tag{13.20}$$

which are the linear transformation laws that apply close to the fixed point. At the fixed point, designated by *, this becomes

$$u^* = b^2 u^* + C(b^2 - b^\epsilon)\lambda^*, \tag{13.21}$$

$$\lambda^* = b^\epsilon \lambda^*, \tag{13.22}$$

which, with $b^2, b^\varepsilon > 1$, can only be satisfied by requiring

$$u^* = \lambda^* = 0. \tag{13.23}$$

The coefficients in Eqs. (13.19) and (13.20) represent the entries to the operator $\hat{\mathcal{M}}^{(b)}$ of Eqs. (9.4)–(9.6), which in the present case specify the linear departures from the (vanishing) fixed point, $u' = u^* = 0$, $\lambda' = \lambda^* = 0$. The matrix for this case reads

$$\hat{\mathcal{M}}^{(b)} = \begin{bmatrix} b^2 & C(b^2 - b^\epsilon) \\ 0 & b^\epsilon \end{bmatrix}, \tag{13.24}$$

which has the form derived in Eq. (9.37), with the corresponding determinant

$$\mathfrak{A}_b = \begin{vmatrix} b^2 & C(b^2 - b^\epsilon) \\ 0 & b^\epsilon \end{vmatrix} = 0. \tag{13.25}$$

Standard techniques, described in the Appendix to Chapter 9, yield the following eigenvalues and eigenvectors (T denotes the transpose):

- $\lambda_1 = b^2$ and $X_1^{\mathrm{T}} = (1, 0)$.
  Comparison with Eq. (9.14) shows that

$$y_1 = 2.$$

- $\lambda_2 = b^{\epsilon}$ and $X_2^{\mathrm{T}} = (-C, 1)$.
  Comparison with Eq. (9.14) shows that

$$y_2 = \epsilon. \tag{13.26}$$

As shown earlier, at a critical point, y_1 and y_2 should differ in sign, with one relevant and one irrelevant parameter set. This is the case here; we have $\epsilon < 0$, that is, $d > 4$, so that the critical exponents, after all this work, are the same as in mean field theory, namely $\nu = \frac{1}{y_1} = 1/2$ and $\eta = 2d_{\varphi} - d + 2 = 0$. We reach the conclusion that, for $d > 4$, mean field theory applies.

13.5 CASE $d < 4$

Now we consider the case that is more relevant to our real world of three dimensions. Repeating the previous analysis, we note that presently $\epsilon > 0$ and $y_2 > 0$. Thus, the fixed point $u^* = \lambda^* = 0$ is repulsive in both directions. To handle this situation, we need to include higher-order terms. This gets to be a somewhat messy procedure; we skip the intermediate steps, simply citing the final results, on which we can then build further. It is found (see Ma[4]) that, to second order, the renormalization of u and λ take the forms

$$u' = b^2 \left[u + \frac{\lambda}{16\pi^2} \left(\frac{1}{2}\Lambda^2 \left\{ 1 - \frac{1}{b^2} \right\} - u \ln b \right) \right] \tag{13.27}$$

and

$$\lambda' = b^{\epsilon} \left[\lambda - \frac{3\lambda^2}{16\pi^2} \ln b \right] \quad (\epsilon \equiv 4 - d). \tag{13.28}$$

It should be carefully noted that the above relation holds only for "small" values of ϵ. Of course, we do not actually have the luxury of allowing the dimensionality of our real world to deviate only slightly from four. However, the present analysis offers some valuable insights. Also, a finite renormalization process requires the repetitive step operation of the above relations.

We begin with small advances in the scaling parameter b by introducing a parameter ρ such that

$$\ln b = \ln(1 + \rho) \approx \rho, \tag{13.29}$$

4. S.-k. Ma, *Modern Theory of Critical Phenomena* (Westview Press, Oxford, 2000), Chapter 7.

in terms of which, keeping only first-order terms in ρ, we reformulate (13.27) as

$$u' = (1+2\rho)\left[u + \frac{\lambda}{16\pi^2}\left(\frac{1}{2}\Lambda^2\left\{1 - \frac{1}{(1+2\rho)^2}\right\}\right) - \frac{\lambda u}{16\pi^2}\rho\right], \qquad (13.30)$$

which may be rewritten in the form

$$\frac{u'-u}{\rho} = 2u + \frac{\Lambda^2\lambda}{16\pi^2} - \frac{\lambda u}{16\pi^2}. \qquad (13.31)$$

In the limit as $\rho \to 0$, the left-hand side becomes a derivative, namely,

$$\frac{du(b)}{d\ln b} = 2u(b) + \frac{\Lambda^2\lambda(b)}{16\pi^2} - \frac{\lambda(b)u(b)}{16\pi^2}. \qquad (13.32)$$

In similar fashion, starting with (13.28), we write

$$\lambda' - \lambda = -\lambda(1-b^{\epsilon}) - \frac{3\lambda^2}{16\pi^2}\ln b = \lambda\epsilon\ln b - \frac{3\lambda^2}{16\pi^2}\ln b, \qquad (13.33)$$

where we had introduced the relation $\ln x \approx x - 1$ for $0 < x < 2$. In the limit as $\rho \to 0$, this leads directly to the derivative form

$$\frac{d\lambda(b)}{d\ln b} = \epsilon\lambda(b) - \frac{3\lambda^2(b)}{16\pi^2}. \qquad (13.34)$$

Eqs. (13.32) and (13.34) allow us to determine the fixed point, based on the condition

$$\left[\frac{du(b)}{d\ln b}\right]_{u=u^*} = \left[\frac{d\lambda(b)}{d\ln b}\right]_{\lambda=\lambda^*} = 0. \qquad (13.35)$$

It is conventional to introduce the symbols $\beta_u \equiv -\frac{du(b)}{d\ln b}$ and $\beta_\lambda \equiv -\frac{d\lambda(b)}{d\ln b}$ and, more generally, $\beta_q(K_q(b)) \equiv -\frac{dK_q(b)}{d\ln b}$ for any coupling constant K_q. Then, at the fixed point,

$$\beta_q(K_q^*) = 0. \qquad (13.36)$$

These relations are examples of the so-called *Callan–Symanzik beta functions*, which play an important role in our further development.

13.6 DETERMINATION OF THE FIXED POINT

We first impose the requirement Eq. (13.35) on Eq. (13.34), yielding the solutions

$$\lambda^* = 0 \text{ and } \lambda^* = 16\frac{\pi^2\epsilon}{3}, \qquad (13.37)$$

of which the second relation is nontrivial.

Dealing with Eq. (13.32), we neglect the last term as being of higher order; imposing requirement (13.35) and adopting the nontrivial solution for λ^*, we obtain

$$u^* = -\frac{\Lambda^2 \epsilon}{6}. \tag{13.38}$$

Notice again the dependence on the dimensionality of the system and the manner in which the cutoff parameter Λ makes its appearance. We have now successfully specified the critical conditions that prevail in the current approximation.

We conclude by studying the linear deviation away from the fixed point, which proceeds as follows: set $\frac{3\lambda^*}{16\pi^2} = \epsilon$; then

$$\frac{d\lambda(b)}{d\ln b} = -\epsilon(\lambda - \lambda^*), \tag{13.39}$$

and

$$\frac{du(b)}{d\ln b} = \left(2 - \frac{\epsilon}{3}\right)\left(u - u^*\right) + \frac{\Lambda^2}{16\pi^2}\left(1 + \frac{\epsilon}{6}\right)\left(\lambda - \lambda^*\right). \tag{13.40}$$

Now go back to Eq. (9.14) for the definitions of y_1 and y_2. These involve the linear departures from the fixed point: y_1 relates to the temperature-dependent phase changes and hence u; y_2 refers to the remaining approach to the fixed point. Thus, from the first term in (13.40) and from (13.39) we obtain

$$y_1 = 2 - \frac{\epsilon}{3} > 0; \quad y_2 = -\epsilon < 0 \quad \text{for } d < 4. \tag{13.41}$$

This thus represents a mixed fixed point that governs the critical behavior in this approximation. Then

$$\nu = \frac{1}{y_1} \approx \frac{1}{2} + \frac{\epsilon}{12}, \tag{13.42}$$

whereas the earlier analysis with respect to η still applies, so that $\eta = 0$.

The other macroscopic coefficients may now be determined. Based on Eqs. (10.33) and setting $d = 4 - \epsilon$, we obtain in first-order approximation

$$\alpha = 2 - \frac{d}{y_1} = 2 - \frac{4-\epsilon}{2 - \frac{\epsilon}{3}} \approx \frac{\epsilon}{6}, \tag{13.43}$$

$$\beta = \frac{1}{2} - \frac{\epsilon}{6}, \tag{13.44}$$

$$\gamma = 2\nu = 1 + \frac{\epsilon}{6}, \tag{13.45}$$

$$\delta = 3 + \epsilon. \tag{13.46}$$

TABLE 13.1 Tabulation of critical exponents as obtained under a variety of models

Exponents	Mean field theory	Renormalized present section	Numerical results for Ising model with $d = 3$	Experimental average value
α	0	0.17	0.12	0.11
β	0.5	0.33	0.31	0.32
γ	1	1.17	1.25	1.24
δ	3	4	5.20	4.75
ν	0.5	0.58	0.64	0.63
η	0	0	0.06	0.03

Let us assess where we now stand. Eqs. (13.43)–(13.46) specify various critical exponents in the limit of small ϵ. We therefore expect these relations to apply roughly in the range $0 < \epsilon < 0.1$. We nevertheless can ask what happens if we set $\epsilon = 1$, so as to be applicable to our real world. The resulting findings are displayed in Table 13.1, which also shows the values corresponding to the mean field approach and to the values obtained by a numerical simulation of the Ising model for three dimensions. We see that the values (13.43)–(13.46) represent a significant improvement over the mean field parameters, being (with one exception) in surprisingly reasonable agreement with those of the presumably exact numerical calculations and with experimental results.[5]

Although we have achieved a significant general improvement over the mean field theory, there still exists a nontrivial discrepancy in the calculated vs. the experimental value for δ. Also, in common with mean field methodology, the predicted η value is zero, in disagreement with the small positive quantity in the range $0.01 \leq \eta \leq 0.06$ encountered experimentally. Attaining η values in the indicated range is no mean feat.

13.7 CASE $d = 4$

There remains the third possibility, that of having $\epsilon = 0$, the so-called *marginal case*. The corresponding treatment is beyond the purview of the present approach; you are encouraged to consult standard monographs for information.

5. This is by no means the only instance where, going well beyond the range of applicability of a first-order approximation, the produced results are in reasonable agreement with experiment. Examples that come to mind include Ohm's law, Fick's law of diffusion, or thermoelectric effects, such as the Seebeck coefficient.

APPENDIX 13.A

Here we provide a cursory justification for relation (13.1). We return to the discretized model of Chapter 11 and focus on an RGT operation that results in the creation of a particular block structure, characterized by a magnetization $M(\boldsymbol{r})$; as earlier, we assume M to vary gently as one passes from one block to the next. The probability of encountering the block structure under study is proportional to $\mathcal{P} \sim \exp[-H'\mathfrak{S}'_\alpha]$. In the original Ising spin lattice, there are many microscopic spin configurations $\mathfrak{S}_i$ consistent with that particular magnetization profile. The sum over all such microstates, giving rise to the partition function, is then related to $\mathcal{P}$ in the following manner:

$$e^{-\hat{\mathcal{H}}'\mathfrak{S}'_\alpha} = \sum_{\mathfrak{S}_i} e^{-\hat{\mathcal{H}}\mathfrak{S}_i}. \tag{13.A.1}$$

In fact, this summation involves all microscopic configurations relating to the coarse-grained order parameter $M(\boldsymbol{r})$, which varies smoothly with position. This allows us to shift from the discretized to a continuum representation of the partition function, where $\varphi(\boldsymbol{r})$ takes the place of $M \equiv \eta$.

We now follow the discussion in Appendix 12.C, Chapter 12. In summation (13.A.1), the discretized spin variables will be supplanted by a catalogue of order parameter functions $\varphi(\boldsymbol{r})$, properly summed via the functional integral

$$e^{-\hat{\mathcal{H}}'(\varphi'(\boldsymbol{r}))} = \int \mathcal{D}\varphi(\boldsymbol{r}) e^{-\hat{\mathcal{H}}(\varphi(\boldsymbol{r}))}. \tag{13.A.2}$$

More explicitly, we may specify the partition function as

$$\mathcal{Z} = \int \mathcal{D}\varphi(\boldsymbol{r}) \exp\left\{-\int d^d\boldsymbol{r}\left[\frac{1}{2}(\nabla\varphi)^2 + \frac{1}{2}u\varphi^2 + \frac{\lambda}{4!}\varphi^4 - \mu\mathcal{B}\varphi\right]\right\}. \tag{13.A.3}$$

This agrees with Eq. (11.6) with $\alpha = 1$.

Setting $B = 0$ and dividing φ into φ^+ and φ^-, we recover Eq. (13.4), with $\hat{\mathcal{H}}$ specified by Eq. (13.1).

Chapter 14

Correlation Functions

We now turn to a study of correlation functions, especially those close to the critical point of a system, which are basic to a characterization of long-range correlations. We review some fundamentals, beginning with a review and generalization of the discussion of Chapter 3.

14.1 CORRELATIONS INVOLVING LATTICE SITES

Consider a system consisting of an array of sites i whose microvariables are specified by $S_1, S_2, \ldots, S_N$. For definiteness, we may refer to these as "spins," but the discussion is perfectly general.

In the simplest case, we deal with the energetics of a system in a field-free region and seek to find the expectation values for a particular variable X_i in that system. One possible approach involves introducing into the standard partition function $\mathcal{Z} = \sum_q e^{-\beta E_q}$ a constant "fictitious field" Y along with a conjugate X_i in the manner shown below, and then discarding Y in the last step of the operation. The augmented partition function in this case takes the form

$$\mathcal{Z} = \sum_q e^{-\beta(E_q - X_i Y)}, \tag{14.1}$$

on which we then carry out the sequence of operations

$$\begin{aligned}\frac{1}{\beta}\left(\frac{\partial \ln \mathcal{Z}}{\partial Y}\right)_{Y=0} &= \frac{1}{\beta \mathcal{Z}}\frac{\partial}{\partial Y}\left(\sum_q e^{-\beta(E_q - X_i Y)}\right)_{Y=0} \\ &= \frac{1}{\mathcal{Z}}\left(\sum_q X_i e^{-\beta(E_q - X_i Y)}\right)_{Y=0} = \langle X_i \rangle,\end{aligned} \tag{14.2}$$

which provides the desired result. Obviously, this method should be viewed as a mathematical "trick," not as a record of actual events.

We next consider a set of spins subjected to a constant magnetic field B. Then, in addition to the ordinary energy states ε_q in the absence of the field, the Hamiltonian contains the contribution

$$\Delta\varepsilon = -\mu B \sum_k S_k. \tag{14.3}$$

A Primer to the Theory of Critical Phenomena. http://dx.doi.org/10.1016/B978-0-12-804685-2.00014-0

Operation (14.2) then allows us determine the magnetization (density) of the N sites as

$$\begin{aligned}\frac{1}{\beta N}\left(\frac{\partial \ln \mathcal{Z}}{\partial \mu B}\right) &= \frac{1}{\beta\mu N\mathcal{Z}}\frac{\partial}{\partial B}\left(\sum_q e^{-\beta(\varepsilon_q-\mu B\sum_k S_k)}\right) \\ &= \frac{1}{N\mathcal{Z}}\sum_q\sum_k S_k e^{-\beta(\varepsilon_q-\mu B\sum_k S_k)} = \frac{1}{N}\sum_k \langle S_k\rangle = M.\end{aligned} \tag{14.4}$$

We now generalize the model by allowing a superposed (magnetic) field $B = B_k$ to vary with position k. Then, in addition to the ordinary energy states ε_q of the various sites in the absence of the external field, the Hamiltonian contains the contribution

$$\Delta\varepsilon = -\frac{1}{\beta}\sum_k J_k S_k, \tag{14.5}$$

where J_k is the usual field variable at $k : J_k = \beta B_k$; to lighten the notation a bit, if B_k refers to a magnetic field, then we also set $\mu = 1$, which should be kept in mind. The resulting partition function of the system in the applied field is written out as

$$\mathcal{Z}_S = \sum_q e^{-(\beta\varepsilon_q-\sum_k J_k S_k)}. \tag{14.6}$$

We can utilize $\mathcal{Z}_S$ to determine the expectation value $\langle S_i\rangle$ of a typical spin by carrying out the following sequence of operations:

$$\begin{aligned}\frac{1}{\mathcal{Z}_S}\left(\frac{\partial \mathcal{Z}_S}{\partial J_i}\right) &= \frac{1}{\mathcal{Z}_S}\frac{\partial}{\partial J_i}\sum_q e^{-(\beta\varepsilon_q-\sum_k J_k S_k)} = \frac{1}{\mathcal{Z}_S}\sum_q S_i e^{-(\beta\varepsilon_q-\sum_k J_k S_k)} \\ &= \frac{1}{\mathcal{Z}_S\beta}\left(\frac{\partial \mathcal{Z}_S}{\partial B_i}\right) = \langle S_i\rangle.\end{aligned} \tag{14.7}$$

This is most easily checked by writing out a few terms of the partition function and then executing the indicated operations.

Proceeding with a second differentiation of Eq. (14.7) for a specific i and l, we find that

$$\begin{aligned}\frac{1}{\mathcal{Z}_S}\left(\frac{\partial^2 \mathcal{Z}_S}{\partial J_i\partial J_l}\right) &= \frac{1}{\mathcal{Z}_S}\sum_q S_i S_l e^{-(\beta\varepsilon_q-\sum_k J_k S_k)} = \frac{1}{\mathcal{Z}_S\beta^2}\left(\frac{\partial^2 \mathcal{Z}_S}{\partial B_i\,\partial B_l}\right) \\ &= \langle S_i S_l\rangle \equiv G(i,l),\end{aligned} \tag{14.8}$$

a pattern that is followed for the higher derivatives. This process thus leads directly to *correlation functions*, as designated by the symbol on the right. The homogeneous field counterpart has the form

$$\frac{1}{\mathcal{Z}_S}\left(\frac{\partial^2 \mathcal{Z}_S}{\partial J^2}\right)=\sum_{i,j}\langle S_i S_j\rangle. \tag{14.9}$$

Thus, the various derivatives of $\ln \mathcal{Z}$ or $\ln \mathcal{Z}_S$ are closely linked to correlation functions. Their utility may be illustrated by obtaining the isothermal static magnetic susceptibility density through differentiations of the magnetization with respect to the uniform (magnetic) field $J_i = \beta B$ for all i. Thus, using Eqs. (14.4), (14.7), and (14.9), we obtain

$$\begin{aligned}\chi_T &= \left(\frac{\partial M}{\partial B}\right)_{B=0} = \frac{1}{N\beta}\frac{\partial}{\partial B}\left[\frac{1}{\mathcal{Z}_S}\left(\frac{\partial \mathcal{Z}_S}{\partial B}\right)\right]_{B=0}\\ &= \frac{k_B T}{N}\left[\frac{1}{\mathcal{Z}_S}\frac{\partial^2 \mathcal{Z}_S}{\partial B^2}-\frac{1}{\mathcal{Z}_S^2}\left(\frac{\partial \mathcal{Z}_S}{\partial B}\right)^2\right]_{B=0}\\ &= (Nk_B T)^{-1}\left[\sum_{i,j}\langle S_i S_j\rangle-\left(\sum_i\langle S_i\rangle\right)^2\right]=\chi_T(i,j)\,.\end{aligned} \tag{14.10}$$

As before, the experimentally determined observable χ_T is linked to spin correlation and expectation values.

A different type of correlation is encountered through the following sequence of operations:

$$\begin{aligned}\frac{\partial}{\partial J_j}\left(\frac{\partial \ln \mathcal{Z}_S}{\partial J_i}\right) &= \frac{\partial}{\partial J_j}\left(\frac{1}{\mathcal{Z}_S}\frac{\partial \mathcal{Z}_S}{\partial J_i}\right)=\frac{1}{\mathcal{Z}_S}\left(\frac{\partial^2 \mathcal{Z}_S}{\partial J_i J_j}\right)-\frac{1}{\mathcal{Z}_S^2}\left(\frac{\partial \mathcal{Z}_S}{\partial J_l}\right)\left(\frac{\partial \mathcal{Z}_S}{\partial J_i}\right)\\ &= \langle S_i S_j\rangle-\langle S_i\rangle\langle S_j\rangle\equiv G_c(i,j)\,.\end{aligned} \tag{14.11}$$

The last symbol represents the *connected correlation function*. It relates to the correlations over and above those expected for noninteracting pairs.

14.2 CORRELATIONS IN THE CONTINUUM LIMIT

We now go over to the continuum case, in which the discrete variables S_i and S_j are replaced by *order parameters* $\varphi(\boldsymbol{r})$ and $\varphi(\boldsymbol{r}')$ that vary continuously with position $\boldsymbol{r}$.

By a generalization of the above presentation we argue that Eqs. (14.10) and (14.11) morph into the following continuum counterpart:

$$\begin{aligned}\chi_T\left(\boldsymbol{r},\boldsymbol{r}'\right) &= \frac{1}{\beta}\left\{\frac{1}{\mathcal{Z}_S}\frac{\partial^2 \mathcal{Z}_S}{\partial B(\boldsymbol{r})\partial B(\boldsymbol{r}')} - \frac{1}{\mathcal{Z}_S}\frac{\partial \mathcal{Z}_S}{\partial B(\boldsymbol{r})}\cdot\frac{1}{\mathcal{Z}_S}\frac{\partial \mathcal{Z}_S}{\partial B(\boldsymbol{r}')}\right\} \\ &= \beta\left\{\langle\varphi\left(\boldsymbol{r}\right)\varphi\left(\boldsymbol{r}'\right)\rangle - \langle\varphi\left(\boldsymbol{r}\right)\rangle\langle\varphi\left(\boldsymbol{r}'\right)\rangle\right\} \equiv \beta G_c\left(\boldsymbol{r},\boldsymbol{r}'\right).\end{aligned} \tag{14.12}$$

This is an example of *linear response theory*, here relating the isothermal susceptibility, a response to the application of a field, to the two-point connected correlation function in the direct space continuum. In most cases, we can replace $\chi_T\left(\boldsymbol{r},\boldsymbol{r}'\right)$ by $\chi_T\left(\left|\boldsymbol{r}-\boldsymbol{r}'\right|\right)$ and $G_c\left(\boldsymbol{r},\boldsymbol{r}'\right)$ by $G_c\left(\left|\boldsymbol{r}-\boldsymbol{r}'\right|\right)$, so that these functions depend only on the separation distances of objects located at points $\boldsymbol{r}$ and $\boldsymbol{r}'$, independent of their relative orientation.

This result may immediately be transferred to reciprocal space via a Fourier transform:

$$\chi_T\left(\boldsymbol{k}\right) = \beta G_c\left(\boldsymbol{k}\right). \tag{14.13}$$

14.3 SPECIFICATION OF THE TWO-POINT CONNECTED CORRELATION FUNCTION

It is of interest to determine the above correlation functions explicitly. For this purpose, we return to the Landau functional, Eq. (11.5), and impose the equilibrium constraint $\frac{\partial \mathcal{L}}{\partial\varphi(\boldsymbol{r})} = 0$. This calls for a functional differentiation whose relevant features are summarized in the Appendix. In many cases, we can simply follow the ordinary derivative operations. Executing those prescriptions, we eliminate the integration step and obtain the equilibrium constraints in the form ($\mu = 1$)

$$-\alpha\nabla^2\varphi\left(\boldsymbol{r}\right) + ut\varphi\left(\boldsymbol{r}\right) + 2b\varphi^3(\boldsymbol{r}) = B(\boldsymbol{r}). \tag{14.14}$$

Here we have explicitly introduced the anticipated linear t dependence of the u parameter that we had introduced earlier, and we have simplified the notation somewhat. The above order parameter function $\varphi(\boldsymbol{r})$ prevails at equilibrium.

Next, take the functional derivative with respect to $B(\boldsymbol{r}')$; this step introduces the Dirac delta distribution on the right. On the left, we introduce a generalization of the static magnetic susceptibility (density): we define, with $\mu = 1$,

$$\chi_T\left(\boldsymbol{r}-\boldsymbol{r}'\right) = \left[\frac{\partial\langle\varphi\left(\boldsymbol{r}\right)\rangle}{\partial B(\boldsymbol{r}')}\right]_{B=0}. \tag{14.15}$$

This leads to

$$\left[-\alpha\nabla^2 + ut + 6b\varphi^2(\boldsymbol{r})\right]\chi_T\left(\boldsymbol{r}-\boldsymbol{r}'\right) = \delta\left(\boldsymbol{r}-\boldsymbol{r}'\right). \tag{14.16}$$

In view of Eq. (14.12), we end up with the following differential equation for $G_c(\boldsymbol{r},\boldsymbol{r}')$:

$$\beta\left[-\alpha\nabla^2 + ut + 6b\varphi^2(\boldsymbol{r})\right]G_c\left(\boldsymbol{r}-\boldsymbol{r}'\right) = \delta\left(\boldsymbol{r}-\boldsymbol{r}'\right), \tag{14.17}$$

showing, on account of the Dirac delta on the right, that G_c is actually a Green function.

There are now two cases to consider:

(i) $t \equiv (T - T_c)/T_c > 0$; we are above the critical temperature, whence the order parameter φ vanishes; then the equation to be solved has the form

$$\left[-\nabla^2 + \xi_u^{-2}\right]G_c\left(\boldsymbol{r}-\boldsymbol{r}'\right) = \frac{k_B T}{\alpha}\delta\left(\boldsymbol{r}-\boldsymbol{r}'\right); \quad \xi_u \equiv \left(\frac{\alpha}{ut}\right)^{\frac{1}{2}}. \tag{14.18}$$

(ii) $t < 0$; we are below the critical temperature, so that we have the solution $\varphi = (-ut/2b)^{1/2}$; then,

$$\left[-\nabla^2 + \xi_l^{-2}\right]G_c\left(\boldsymbol{r}-\boldsymbol{r}'\right) = \frac{k_B T}{\alpha}\delta\left(\boldsymbol{r}-\boldsymbol{r}'\right); \quad \xi_l \equiv \left(-\frac{\alpha}{2ut}\right)^{\frac{1}{2}}. \tag{14.19}$$

Both cases are subsumed in the relation

$$\left[-\nabla^2 + \xi^{-2}\right]G_c\left(\boldsymbol{r}-\boldsymbol{r}'\right) = \frac{k_B T}{\alpha}\delta\left(\boldsymbol{r}-\boldsymbol{r}'\right). \tag{14.20}$$

As we demonstrate below, ξ, with dimension of distance, is the correlation length.

The reciprocal space version $G_c(\boldsymbol{k})$ of the two-point correlation function is obtained by taking the Fourier transform of Eq. (14.20): following standard methodology, the Fourier transform of the second derivative on the left gives rise to the factor $k^2 G_c(\boldsymbol{k})$, which is then followed by $\xi^{-2}G_c(\boldsymbol{k})$; on the right the Fourier transform of the delta "function," including the multiplier, yields $\frac{k_B T}{\alpha}$. Thus,

$$G_c(\boldsymbol{k}) = \frac{k_B T}{\alpha}\frac{1}{k^2 + \xi^{-2}}. \tag{14.21}$$

At $T = T_c$, $\xi \to \infty$ and $G_c(\boldsymbol{k}) \propto k^{-2}$.

However, we are also interested in the case $T \to T_c$. Rewrite Eq. (14.21) as

$$G_c(k) = \frac{k_B T}{\alpha}\frac{\xi^2}{1 + k^2\xi^2}. \tag{14.22}$$

By introducing the isothermal susceptibility, Eq. (14.13), and setting $k = 0$, we find that

$$G_c(0) = \frac{k_B T}{\alpha}\xi^2 = k_B T \chi_T(0), \tag{14.23}$$

whence

$$\chi_T(0) = \frac{\xi^2}{\alpha}. \tag{14.24}$$

Eliminating α between (14.22) and (14.23), we obtain

$$G_c(\boldsymbol{k}) = \frac{k_B T \chi_T(0)}{1 + k^2 \xi^2}, \tag{14.25}$$

which is the final formulation for the connected correlation functions in reciprocal space.

14.4 CORRELATIONS IN DIRECT SPACE

To find the direct space version requires more work. We begin with the case of three dimensions. Start with Eq. (14.21) and execute the inverse Fourier transform to the direct space:

$$\frac{G_c(\boldsymbol{r})}{k_B T/\alpha} = \frac{1}{(2\pi)^3} \int d\boldsymbol{k}^3 \frac{e^{-i\boldsymbol{k}\cdot\boldsymbol{r}}}{k^2 + \xi^{-2}}. \tag{14.26}$$

The integration is executed in detail in the Appendix. We find that

$$\frac{G_c(\boldsymbol{r})}{k_B T/\alpha} = \frac{1}{4\pi r} e^{-r/\xi}, \tag{14.27}$$

which also shows that ξ is to be interpreted as a correlation distance. Notice the screened Coulomb potential variation.

The extension to different dimensions is more troublesome, requiring the solution of Eq. (14.20). Without loss of generality, we may set $\boldsymbol{r}' = 0$; also, for the spherically symmetric case, we may introduce the radial polar coordinate system in d dimensions: Eq. (14.20) then reads

$$\left[-\frac{1}{r^{d-1}} \frac{\partial}{\partial r} r^{d-1} \frac{\partial}{\partial r} + \xi^{-2}\right] G_c(r) = \frac{k_B T}{\alpha} \delta(r). \tag{14.28}$$

Notice how the dimensionality of the system enters at this point. This equation is well known in applied mathematics; the d-dependent solutions involve modified spherical Bessel functions of the third kind. Instead of displaying all the formidable formulas, it makes more sense first to show in Fig. 14.1 a semilogarithmic plot of scaled connected correlation functions G_c vs. the scaled variable $\frac{r}{\xi}$ for various d. The denominator in the abscissa label is the G_c function for $r = \xi$. Note the dependence on d and the anticipated very steep rise in the connected correlation functions with decreasing $\frac{r}{\xi}$.

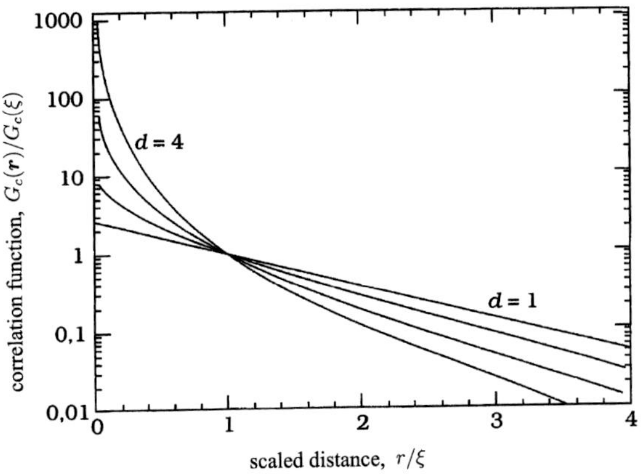

FIGURE 14.1 Connected correlation functions in direct space in various dimensions d. Adapted from Binney et al., *The Theory of Critical Phenomena* (Oxford University Press, 2002, p. 194).

It is, however, important to present formulas that apply in certain limiting cases. In the solution for $d \geq 2$, we distinguish between two limits:

When $T \neq T_c$, $r \gg \xi$:

$$G_c(r) = \frac{\pi^{(1-d)/2}}{\pi^{(1+d)/2}} \cdot \frac{k_B T}{\alpha} \cdot \frac{1}{\xi^{(d-3)/2}} \cdot \frac{e^{-r/\xi}}{r^{(d-1)/2}}, \quad G_c \sim e^{-r/\xi}. \tag{14.29}$$

Note again the dependence of G_c on distance, dominated by the exponential factor. Also involved is the dependence of the correlation length, and of the position variable, on the dimension of the system.

On the other hand, when $T = T_c$, the correlation length $\xi \to \infty$, and the solution approaches the opposite asymptotic case, $r \ll \xi$: namely for $d > 2$,

$$G_c(\boldsymbol{r}) = \frac{\Gamma\left(\frac{d-2}{2}\right) k_B T}{4\pi^{d/2}\alpha} \cdot \frac{1}{r^{d-2}}; \quad G_c \sim \frac{1}{r^{d-2}} \tag{14.30}$$

(Γ is the standard gamma function), where the distance dependence is now algebraic and again involves the dimension d. However, in detailed experimental investigations we encounter a power law decay of the form

$$G_c(r) \sim \frac{A}{r^{d-2+\eta}}, \tag{14.31}$$

with η (which is standard notation, not to be confused with the order parameter symbol we used earlier) in the range 0.016 to 0.06, which is "small" but not negligible. A resolution of this disparity turns out to require considerable effort.

APPENDIX 14.A FUNCTIONAL DERIVATIVES

We provide a highly cursory description of functional derivatives that are extensively used in the presentation of field theories. We amplify here the discussion of Appendix 12.D. Let $F[\phi(x)]$ be a functional (i.e., a function of a function) that involves an ordinary function $\phi(x)$ that has x as an independent variable. Then the change δF of the functional arising from a variation $\delta\phi(x)$ is specified by the defining relation

$$\delta F[\phi] = \int dx \frac{\partial F[\phi]}{\partial \phi(x)} \delta\phi(x), \tag{14.A.1}$$

which states that the total change in F involves a linear superposition of all local F changes due to $\delta\phi(x)$ summed over all x values. The above accords with the ordinary definition of a derivative if we specify the local variation in $\phi(x)$ as

$$\delta\phi(x) = \varepsilon\delta(x-y), \tag{14.A.2}$$

where ε is the magnitude of the change in ϕ at position y. We can then write

$$\delta F[\phi] = F\left[\phi + \varepsilon\delta(x-y)\right] - F[\phi] = \int dx \frac{\partial F[\phi]}{\partial \phi(x)} \varepsilon\delta(x-y) = \varepsilon \frac{\partial F[\phi]}{\partial \phi(y)}, \tag{14.A.3}$$

which may be recast in the more familiar form as

$$\frac{\partial F[\phi]}{\partial \phi(y)} = \lim_{\varepsilon\to 0} \frac{1}{\varepsilon} \left\{F\left[\phi(x) + \varepsilon\delta(x-y)\right] - F[\phi(x)]\right\}. \tag{14.A.4}$$

This finds its application in the study of functionals. Consider the object $F[G[\phi]]$. Then applying the chain rule of differentiation to Eq. (14.A.1), we find that

$$\frac{\partial}{\partial \phi(y)} F[G[\phi]] = \int dx \frac{\partial F[G]}{\partial G[\phi(x)]} \frac{\partial G[\phi]}{\partial \phi(y)}. \tag{14.A.5}$$

As a particular case, we may take G to be an ordinary function $g(\phi)$ that depends on x. Then the integral is wiped out, and

$$\frac{\partial}{\partial \phi(y)} F[g(\phi)] = \frac{\partial F}{\partial g(\phi(y))} \frac{\partial g(\phi)}{\partial \phi(y)}. \tag{14.A.6}$$

As a particular example, consider $F[\phi] = \int dx\, [\phi(x)]^n$. Applying Eq. (14.A.4), we find that

$$\begin{aligned}\frac{\partial F[\phi]}{\partial \phi(y)} &= \lim_{\varepsilon \to 0} \frac{1}{\varepsilon} \left\{ \int dx \left[\phi(x) + \varepsilon \delta(x-y)\right]^n - \int dx\, [\phi(x)]^n \right\} \\ &= \int dx n\, [\phi(x)]^{n-1}\, \delta(x-y) = n\left[\phi(y)\right]^{n-1}.\end{aligned} \tag{14.A.7}$$

Note that we have eliminated the integration over x and executed a normal differentiation of the integrand. This result finds its direct application in the present section.

You should also be able to verify the following relations:

$$\frac{\delta}{\delta \phi(y)} \int dx \phi(x) = 1, \tag{14.A.8}$$

$$\frac{\delta}{\delta \phi(y)} \phi(x) = \delta(x-y). \tag{14.A.9}$$

Finally, you may show that

$$\frac{\delta}{\delta \phi(y)} \left\{ \frac{1}{2} \int d^d x\, [\nabla \phi(x)]^2 \right\} = -\nabla^2 \phi(y). \tag{14.A.10}$$

APPENDIX 14.B EVALUATION OF CORRELATION FUNCTION IN DIRECT SPACE FOR THREE DIMENSIONS

For the determination of $G_c(r)$ in three dimensions, we begin by taking the Fourier transform of Eq. (14.26):

$$\frac{G_c(\boldsymbol{r})}{k_B T/\alpha} = \frac{1}{(2\pi)^3} \int d^3 \boldsymbol{k} \frac{e^{-ikr\cos\theta_k}}{k^2 + \xi^{-2}}. \tag{14.B.1}$$

With $d^3\boldsymbol{k} = k^2 dk \sin\theta_k d\theta_k d\varphi_k$, the integration over the azimuth angle φ_k yields a factor of 2π; the polar angle integration involves setting $x = \cos\theta_k$, $dx = -\sin\theta_k d\theta_k$, whence

$$\int_{-\frac{\pi}{2}}^{\frac{\pi}{2}} \exp(-ikr\cos\theta_k) \sin\theta_k d\theta_k = -\int_1^{-1} e^{-ikrx} dx = \frac{e^{ikr} - e^{-ikr}}{ikr} = \frac{2\sin kr}{kr}, \tag{14.B.2}$$

so that

$$\frac{G_c(\boldsymbol{r})}{k_B T/\alpha} = \frac{1}{2\pi^2 r} \int_0^\infty \frac{kdk \sin(kr)}{k^2 + \xi^{-2}} = \frac{1}{4\pi^2 r} \operatorname{Im} \int_{-\infty}^{\infty} \frac{kdke^{ikr}}{k^2 + \xi^{-2}}. \tag{14.B.3}$$

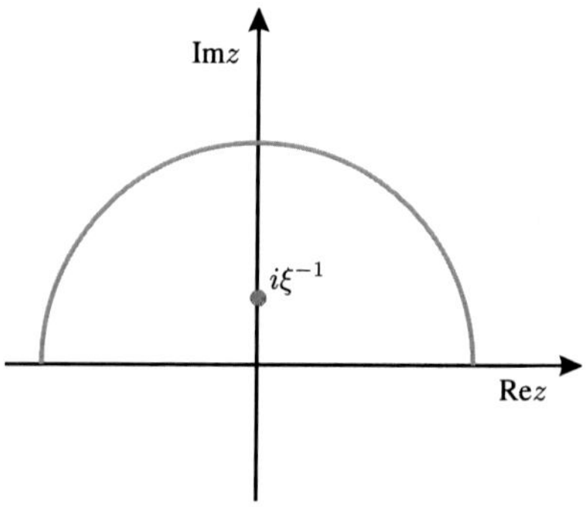

FIGURE 14.2 Contour diagram for the determination of the contour integral in Eq. (14.B.4). See the text for details.

Note the change in sign in the exponent relative to Eq. (14.B.1). Had we chosen $e^{ikr\cos\theta_k}$ as the exponential term, the outcome, Eq. (14.B.3), would have been the same. The choice of $\sqrt{(2\pi)^3}$ or 1^3 as alternative prefactors in Eq. (14.B.1), which is occasionally used, would only affect the ultimate scaling factor, not the distance relation of interest. For reasons shown below, we confine the integration to the range 0 to $-\infty$; it is easy to verify that this incurs an additional factor of $\frac{1}{2}$ without any net sign change in the integral.

Expression (14.B.3) is evaluated by contour integration involving the Cauchy relation

$$\oint \frac{f(z)}{z-z_0}dz = 2\pi i f(z_0). \tag{14.B.4}$$

To render (14.B.4) compatible with (14.B.3), set $z = k^2$ and rewrite the denominator as $k^2 - \left(i\xi^{-1}\right)^2 = z - z_0$, with a pole at $k_0 = \pm i\xi^{-1} = \pm\sqrt{z_0}$. The integration path is shown in Fig. 14.2, being confined to the upper half plane, so that k remains nonnegative. The semicircle has to be extended to an infinite radius, so as to avoid making any contribution to the integral; hence, the limitation of the above-mentioned integration path in Eq. (14.B.3). Then, referring to (14.B.4), we can obtain the two-point correlation function of the form

$$\frac{G_c(\mathbf{r})}{k_B T/\alpha} = \frac{1}{8\pi^2 r}\operatorname{Im}\oint \frac{dz\exp\left(i\sqrt{z}r\right)}{z-z_0} = \frac{1}{4\pi r}e^{-r/\xi}, \tag{14.B.5}$$

where we were obliged to select the positive pole to keep $k \geq 0$. This relation again shows that ξ is to be interpreted as a correlation distance.

Chapter 15

Beyond the Landau Model

We now engage in an initial foray into the more advanced description of critical phenomena that require specialized mathematical techniques and familiarity with field theoretical concepts. We can advance to some degree before having to stop in the face of technical details beyond the purview of this book.

15.1 BASICS

In keeping with the literature on the subject, attention is directed to a convention, introduced in early studies of field theory, of replacing the parameter u in the Hamiltonian by the square of the "bare mass," m_0^2, even when we are not dealing directly with mass concepts and even when m_0^2 is negative. However, we will (reluctantly) temporarily follow this procedure to rederive the Callan–Symanzik equations cited earlier. At this stage, we also introduce the renormalization procedure: basically, we start with a set of noninteracting units, termed the "bare" entities, and then introduce interactions that are represented as "dressed" or renormalized entities. Those familiar with solid-state physics will recognize this procedure: the effect of lattice potentials is simulated by assigning conduction electrons an effective mass different from the free electron mass. Likewise, the effect of all other electrons interacting with a randomly chosen one in a plasma is handled by introducing the concept of a noninteracting quasielectron with an effective charge (and mass) differing from that of a free electron.

The mathematical operations of the last few chapters have already involved "renormalization techniques": a procedure, previously also designated as a rescaling process, whereby the results become independent of the scale of operations. As we approach the critical point, the events initially formulated at atomic scales become irrelevant.

In our previous study of rescaling effects, we achieved the invariance of the Hamiltonian function at the expense of altering the relevant parameters. This alerts us to the fact that proceeding from an initial, "bare" set of parameters, m_0^2 and λ_0, we end up with a "renormalized set" of m^2 and λ. The renormalization also involves the scaling factor b and the dimension parameter $\varepsilon \equiv 4 - d$.

A Primer to the Theory of Critical Phenomena. http://dx.doi.org/10.1016/B978-0-12-804685-2.00015-2

We begin our study with the Hamiltonian density established earlier in terms of the initial order parameter function φ_0:

$$\hbar(\varphi_0) = \frac{1}{2}(\nabla\varphi_0)^2 + \frac{1}{2}m_0^2\varphi_0^2 + \frac{\lambda_0}{4!}\varphi_0^4. \tag{15.1}$$

We can proceed from this "bare" quantity to "dressed" or "exact" quantities by use of the following redefinitions that leave the form of $\hbar(\varphi_0)$ invariant in its transformation to $\hbar(\varphi)$:

$$m^2 = \mathcal{Z}_\varphi m_0^2, \tag{15.2}$$

$$\lambda = \mathcal{Z}_\varphi^2 \lambda_0, \tag{15.3}$$

$$\varphi = \mathcal{Z}_\varphi^{-1/2} \varphi_0, \tag{15.4}$$

where $\mathcal{Z}_\varphi$ is known as the *renormalization constant*. It is easy to check that the proposed changes in variables do leave the Hamiltonian density unchanged, as long as we also require that $(\nabla\varphi_0)^2 = (\nabla\varphi)^2$ remain invariant.

We now introduce the n-particle correlation function, also known as the *bare propagator*,

$$G_0^{(n)}(\boldsymbol{k}_1, \boldsymbol{k}_2, \ldots, \boldsymbol{k}_n) = \langle \varphi_0(\boldsymbol{k}_1), \varphi_0(\boldsymbol{k}_2), \ldots, \varphi_0(\boldsymbol{k}_n) \rangle_0 \tag{15.5}$$

and its exact counterpart as

$$G^{(n)}(\boldsymbol{k}_1, \boldsymbol{k}_2, \ldots, \boldsymbol{k}_n) = \langle \varphi(\boldsymbol{k}_1), \varphi(\boldsymbol{k}_2), \ldots, \varphi(\boldsymbol{k}_n) \rangle. \tag{15.6}$$

We then find that $G^{(n)}$ transforms as the n products of φ, so that

$$G_0^{(n)}(\boldsymbol{k}_1, \boldsymbol{k}_2, \ldots, \boldsymbol{k}_n; \lambda_0, m_0, \varepsilon) = \mathcal{Z}_\varphi^{n/2} G^{(n)}(\boldsymbol{k}_1, \boldsymbol{k}_2, \ldots, \boldsymbol{k}_n, ; \lambda, m, b, \varepsilon). \tag{15.7}$$

It turns out to be more convenient to work with the reciprocal of these propagators. These are the so-called *vertex functions*, which are governed by the relation

$$\Gamma_0^{(n)}(\boldsymbol{k}_1, \boldsymbol{k}_2, \ldots, \boldsymbol{k}_n; \lambda_0, m_0, \varepsilon) = \mathcal{Z}_\varphi^{-n/2} \Gamma^{(n)}(\boldsymbol{k}_1, \boldsymbol{k}_2, \ldots, \boldsymbol{k}_n, ; \lambda, m, b, \varepsilon). \tag{15.8}$$

Here b is the usual scale of renormalization. Again, the subscript zero indicates that we deal with the "free" or noninteracting values of the parameters; they are constants in our subsequent operations.

We now attend to the following operation, noting that $\Gamma_0^{(n)}$ is independent of b:

$$b\frac{d\Gamma_0^{(n)}}{db} = 0 = b\frac{d\left(\mathcal{Z}_\varphi^{-\frac{n}{2}}\Gamma^{(n)}\right)}{db} = b\mathcal{Z}_\varphi^{-\frac{n}{2}}\frac{d\Gamma^{(n)}}{db} + b\Gamma^{(n)}\frac{d\mathcal{Z}_\varphi^{-\frac{n}{2}}}{db}. \tag{15.9}$$

Now apply the chain rule of differentiation; keep in mind that $\mathcal{Z}_\varphi = \mathcal{Z}_\varphi(\lambda, \varepsilon)$ and that $\lambda = \lambda(b)$. The last term in Eq. (15.9) is to be rewritten as shown in the Appendix. We then find that

$$\left[b\frac{\partial}{\partial b} + b\frac{\partial \lambda}{\partial b}\frac{\partial}{\partial \lambda} + b\frac{\partial m}{\partial b}\frac{\partial}{\partial m} - n\gamma_\varphi(\lambda, \varepsilon)\right]\Gamma^{(n)} = 0, \tag{15.10}$$

where

$$\gamma_\varphi(\lambda, \varepsilon) \equiv \mathcal{Z}_\varphi^{-\frac{1}{2}}\left[b\frac{\partial \mathcal{Z}_\varphi^{\frac{1}{2}}(\lambda, \varepsilon)}{\partial b}\right]. \tag{15.11}$$

Eq. (15.10) may be recast in the standard form

$$\left[b\frac{\partial}{\partial b} - \beta(\lambda, \varepsilon)\frac{\partial}{\partial \lambda} - \gamma_m(\lambda, \varepsilon)\frac{\partial}{\partial m} - n\gamma_\varphi(\lambda, \varepsilon)\right]\Gamma^{(n)} = 0 \tag{15.12}$$

with

$$\beta(\lambda, \varepsilon) = -b\frac{\partial \lambda}{\partial b} = -\frac{\partial \lambda}{\partial \ln b}, \tag{15.13}$$

$$\gamma_m(\lambda, \varepsilon) = -b\frac{\partial m}{\partial b} = -\frac{\partial m}{\partial \ln b}. \tag{15.14}$$

Eq. (15.12) is the so-called *Callan–Symanzik equation*, where $\beta(\lambda, \varepsilon)$ is the *Callan–Symanzik beta function* we encountered earlier in a slightly different form, Eq. (13.36). In the Appendix, we show that γ_φ is the anomalous dimension of φ. Also, in positing $m = b^{-\gamma_m} m_0$ for fixed m_0, the derivative $-b\left[\frac{\partial \frac{m}{m_0}}{\partial b}\right]$ generates the function γ_m, whence γ_m represents the anomalous dimension of m.

We now have, as our next objective the expression of $\beta(\lambda, \varepsilon)$ as a power series in the indicated variables.

15.2 THE GENERATING FUNCTIONAL FOR ESTABLISHING CORRELATIONS

We start by examining how the partition function for the Ginzburg–Landau model is set up. This will be needed in our later determination of correlation functions. We begin with the relation

$$\mathcal{Z}[J]=\int \mathcal{D}\varphi e^{-\hat{\mathcal{H}}[J]}=\int \mathcal{D}\varphi e^{-\hat{\mathcal{H}}_0[J]}\exp\left[-\int d^d\boldsymbol{x}\lambda_0\frac{\varphi^4}{4!}\right], \qquad (15.15)$$

where we have split the Gaussian (bare) portion off from the φ^4 contribution. The latter will now be expanded as a power series in λ_0:

$$\mathcal{Z}[J]=\int \mathcal{D}\varphi e^{-\hat{\mathcal{H}}_0[J]}\left[\sum_{n=1}^{\infty}\frac{1}{n!}\left(-\frac{\lambda_0}{4!}\int d^d\boldsymbol{x}\varphi^4\right)^n\right]. \qquad (15.16)$$

As our next step, we exchange the integration and summation operations. This cannot be done with impunity because of convergence problems. However, the series is believed to be asymptotic; that is, a partial summation does approximate the desired result, as long as terms of higher order are ignored. With that in mind, we write

$$\mathcal{Z}[J]=\sum_{n=1}^{\infty}\frac{1}{n!}\int \mathcal{D}\varphi e^{-\hat{\mathcal{H}}_0[J]}\left(-\frac{\lambda_0}{4!}\int d^d\boldsymbol{x}\varphi^4\right)^n, \qquad (15.17)$$

and we multiply the latter by $\frac{\mathcal{Z}_0[J]}{\mathcal{Z}_0[J]}$ to obtain

$$\mathcal{Z}[J]=\mathcal{Z}_0[J]\frac{\sum_{n=1}^{\infty}\frac{1}{n!}\int \mathcal{D}\varphi e^{-\hat{\mathcal{H}}_0[J]}\left(-\frac{\lambda_0}{4!}\int d^d\boldsymbol{x}\varphi^4\right)^n}{\mathcal{Z}_0[J]}. \qquad (15.18)$$

Effectively,

$$\mathcal{Z}[J]=\sum_{n=1}^{\infty}\frac{1}{n!}\left\langle\left(-\frac{\lambda_0}{4!}\int d^d\boldsymbol{x}\varphi^4\right)^n\right\rangle_0\mathcal{Z}_0[J]. \qquad (15.19)$$

Eq. (15.19), involving the angular brackets, is to be interpreted as the average $\langle\varphi^4\rangle_0\equiv G_0^{(4)}$ taken over the Gaussian (not the full Ginzburg–Landau) distribution. Thus, $\mathcal{Z}[J]$ is to be constructed from an asymptotic sum of terms involving the Gaussian correlation function $\langle\varphi^4\rangle_0$. We note that, as explained in Chapter 14, in the absence of an applied external field, we must set $J=0$ at the end.

Following the exposition in Chapter 14, Eqs. (14.A.1)–(14.A.7), we obtain $\langle\varphi^4\rangle_0$ by a fourth-order functional differentiation, whereby Eq. (15.19) becomes

$$\begin{aligned}\mathcal{Z}[J]&=\sum_{n=1}^{\infty}\frac{1}{n!}\left(-\frac{\lambda_0}{4!}\int d^d\boldsymbol{x}\frac{\partial^4}{\partial J^4(\boldsymbol{x})}\right)^n\times\mathcal{Z}_0[J]\Bigg|_{J=0}\\&=\exp\left\{-\frac{\lambda_0}{4!}\int d^d\boldsymbol{x}\frac{\partial^4}{\partial J^4(\boldsymbol{x})}\right\}\times\mathcal{Z}_0[J]\Bigg|_{J=0},\end{aligned} \qquad (15.20)$$

where we reverted back to the exponential form on the second line.

The factor $\mathcal{Z}_0[J]$ in turn can be split into the form shown in Eqs. (12.29) and (12.32) with $\alpha = 1$:

$$\begin{aligned}\mathcal{Z}_0[J] &= \mathcal{Z}_0[0]\exp\left[\frac{1}{2}\int_{0\le k\le\Lambda} d^d\boldsymbol{k}\frac{J(\boldsymbol{k})J(-\boldsymbol{k})}{k^2+u}\right] \\ &= \mathcal{Z}_0[0]\exp\left[\frac{1}{2}\int d^d\boldsymbol{x}\int d^d\boldsymbol{y}\,J(\boldsymbol{x})\Delta(\boldsymbol{x}-\boldsymbol{y})J(\boldsymbol{y})\right].\end{aligned} \tag{15.21}$$

For later convenience, we abbreviate this as

$$\mathcal{Z}_0[J] = \mathcal{Z}_0[0]\exp\left[\frac{1}{2}J\Delta J\right]. \tag{15.22}$$

Combining the above, we obtain the final partition functional in the form

$$\mathcal{Z}[J] = \mathcal{Z}_0[0]\exp\left\{-\frac{\lambda_0}{4!}\int d^d\boldsymbol{x}\frac{\partial^4}{\partial J^4(\boldsymbol{x})}\right\}\exp\left[\frac{1}{2}\int d^d\boldsymbol{x}\int d^d\boldsymbol{y}\,J(\boldsymbol{x})\Delta(\boldsymbol{x}-\boldsymbol{y})J(\boldsymbol{y})\right]. \tag{15.23}$$

15.3 THE TWO-POINT CORRELATION FUNCTION

As mentioned earlier, our principal aim is to determine the correlation functions. We illustrate the procedure by considering the two-point correlation function

$$G^{(2)}(\boldsymbol{x}_1,\boldsymbol{x}_2) \equiv \langle\varphi(\boldsymbol{x}_1)\varphi(\boldsymbol{x}_2)\rangle, \tag{15.24}$$

for which we expand the exponential term in Eq. (15.23). In dealing with these operations, we proceed via the functional differentiation.

We begin with

$$\begin{aligned}G^{(2)}(\boldsymbol{x}_1,\boldsymbol{x}_2) &= \left[\frac{1}{Z[J]}\frac{\partial}{\partial J(\boldsymbol{x}_1)}\frac{\partial}{\partial J(\boldsymbol{x}_2)}\right]\mathcal{Z}[J]\bigg|_{J=0} \\ &= \frac{\mathcal{Z}_0[0]}{Z[J]}\frac{\partial}{\partial J(\boldsymbol{x}_1)}\frac{\partial}{\partial J(\boldsymbol{x}_2)}\exp\left\{-\frac{\lambda_0}{4!}\int d^d\boldsymbol{x}\frac{\partial^4}{\partial J^4(\boldsymbol{x})}\right\}\exp\left[\frac{1}{2}J\Delta J\right]_{J=0}.\end{aligned} \tag{15.25}$$

Consider first the zero-order term in the expansion of the first exponential factor; this is simply unity, so that with $\mathcal{Z}[J] = \mathcal{Z}_0[0]$, we find that

$$\begin{aligned}G^{(2)}(\boldsymbol{x}_1,\boldsymbol{x}_2)\bigg|_0 &= \frac{\partial}{\partial J(\boldsymbol{x}_1)}\frac{\partial}{\partial J(\boldsymbol{x}_2)} \\ &\times\left[\exp\left[\frac{1}{2}\int d^d\boldsymbol{x}\int d^d\boldsymbol{y}\,J(\boldsymbol{x})\Delta(\boldsymbol{x}-\boldsymbol{y})J(\boldsymbol{y})\right]\right]_{J=0}.\end{aligned} \tag{15.26}$$

Now execute the functional differentiation summarized in Section 12.5: we find that

$$
\begin{aligned}
G^{(2)}(\boldsymbol{x}_1,\boldsymbol{x}_2)\Big|_0 &= \left\{\frac{1}{2}\iint d^d\boldsymbol{x}d^d\boldsymbol{y}\frac{\partial J(\boldsymbol{x})}{\partial J(\boldsymbol{x}_1)}\Delta(\boldsymbol{x}-\boldsymbol{y})\frac{\partial J(\boldsymbol{y})}{\partial J(\boldsymbol{x}_2)}\right\}\exp\left[\frac{1}{2}J\Delta J\right]_{J=0} \\
&+\left\{\frac{1}{2}\iint d^d\boldsymbol{x}d^d\boldsymbol{y}\frac{\partial J(\boldsymbol{x})}{\partial J(\boldsymbol{x}_2)}\Delta(\boldsymbol{x}-\boldsymbol{y})\frac{\partial J(\boldsymbol{y})}{\partial J(\boldsymbol{x}_1)}\right\}\exp\left[\frac{1}{2}J\Delta J\right]_{J=0}.
\end{aligned}
\tag{15.27}
$$

Since $\frac{\partial J(\boldsymbol{x})}{\partial J(\boldsymbol{y})} = \delta(\boldsymbol{x}-\boldsymbol{y})$, and because the exponential term is replaced by unity on setting $J=0$, this reduces to

$$
\begin{aligned}
G^{(2)}(\boldsymbol{x}_1,\boldsymbol{x}_2)\Big|_0 &= 2\times\frac{1}{2}\iint d^d\boldsymbol{x}d^d\boldsymbol{y}\delta(\boldsymbol{x}-\boldsymbol{x}_1)\Delta(\boldsymbol{x}-\boldsymbol{y})\delta(\boldsymbol{y}-\boldsymbol{x}_2) \\
&= \Delta(\boldsymbol{x}_1-\boldsymbol{x}_2),
\end{aligned}
\tag{15.28}
$$

which is the expected zeroth-order result.

Next, we should deal with the first-order terms in λ_0. The exfoliation of the mathematical operation represents a perfectly straightforward procedure. Readers interested in the details are referred to the supplementary material in Chapter 17. Here we simply quote the final result, which reads

$$
G^{(2)}(\boldsymbol{x}_1,\boldsymbol{x}_2)\Big|_0 = \Delta(\boldsymbol{x}_1-\boldsymbol{x}_2)+\frac{\lambda_0}{2}\int d^d\boldsymbol{x}\Delta(\boldsymbol{x}_1-\boldsymbol{x})\Delta(\boldsymbol{x}-\boldsymbol{x})\Delta(\boldsymbol{x}-\boldsymbol{x}_2).
\tag{15.29}
$$

This is beginning to take the form of the exact propagator as a power series involving convolutions of bare propagators.

15.4 POSTSCRIPT

To progress beyond this point we need to introduce more advanced mathematical techniques. We are providing a gentle introduction in Chapter 17, to get readers started on a study of the monographs listed at the end of this book.

APPENDIX 15.A ANOMALOUS DIMENSION OF φ

Examine the last term in Eq. (15.9), which involves

$$
\Gamma^{(n)}b\frac{\partial \mathcal{Z}_\varphi^{-\frac{n}{2}}}{\partial b} = -\Gamma^{(n)}\frac{n}{2}\mathcal{Z}_\varphi^{-\frac{n}{2}}\mathcal{Z}_\varphi^{-1}b\frac{\partial \mathcal{Z}_\varphi}{\partial b}.
\tag{15.A.1}
$$

Then notice that

$$\frac{\partial \mathcal{Z}_{\varphi}^{\frac{1}{2}}}{\partial b} = \frac{1}{2}\mathcal{Z}_{\varphi}^{-\frac{1}{2}}\frac{\partial \mathcal{Z}_{\varphi}}{\partial b}. \tag{15.A.2}$$

Elimination of $\frac{\partial \mathcal{Z}_{\varphi}}{\partial b}$ between these relations yields

$$\Gamma^{(n)} b \frac{\partial \mathcal{Z}_{\varphi}^{-\frac{n}{2}}}{\partial b} = -n\mathcal{Z}_{\varphi}^{-\frac{n}{2}}\Gamma^{(n)}\left(\mathcal{Z}_{\varphi}^{-\frac{1}{2}} b \frac{\partial \mathcal{Z}_{\varphi}^{\frac{1}{2}}}{\partial b}\right) \equiv -n\mathcal{Z}_{\varphi}^{-\frac{n}{2}}\Gamma^{(n)}\gamma_{\varphi}(\lambda,\epsilon). \tag{15.A.3}$$

The factor $\mathcal{Z}_{\varphi}^{-\frac{n}{2}}$ is common to all terms in Eq. (15.10) and may therefore be dropped when the last expression is added to Eq. (15.10).

To show that γ_{φ} specifies the anomalous dimension of b, start with the previous relation for γ_{φ}; then from Eq. (15.4) and the definition for the anomalous dimension set we have

$$\frac{\varphi(x)}{\varphi'(x')} = b^{-d_{\varphi}} = \mathcal{Z}_{\varphi}^{\frac{1}{2}}, \quad \frac{\partial \mathcal{Z}_{\varphi}^{\frac{1}{2}}}{\partial b} = -d_{\varphi} b^{-d_{\varphi}-1}, \tag{15.A.4}$$

so that

$$b\left(\mathcal{Z}_{\varphi}^{-\frac{1}{2}}\frac{\partial \mathcal{Z}_{\varphi}^{\frac{1}{2}}}{\partial b}\right) = \gamma_{\varphi} = -d_{\varphi}, \tag{15.A.5}$$

as was to be established.

Chapter 16

An Elementary Examination of Quantum Phase Transitions Involving Fermions*

In this chapter, regarded as supplementary, we turn the reader's attention to the principal extension of the concepts discussed in this book, namely quantum phase transitions, which have taken place during the last 30 years (not counting the pioneering work of John Hertz in 1976). The exposition is largely unsystematic. Nonetheless, we think that even at the primary stage the reader should be aware what has been taking place during those years. The chapter, marked with (*), is meant to imply that the material is not an integral part of the present text. Also, we arbitrarily selected the phase transitions for correlated fermions, as one of the most spectacular phenomena that occur.

16.1 INTRODUCTION: FACTORS DETERMINING A CONTINUOUS QUANTUM PHASE TRANSITION

In the entire book, we have dealt with the equilibrium properties (and associated thermodynamic fluctuations) near *continuous classical phase transitions*, i.e., those with critical temperatures $T_c > 0$. The divergent physical properties such as the specific heat, the magnetic susceptibility, etc., all scale with a divergent correlation length $\xi \sim t^{-\nu}$, with $\nu > 0$ as the corresponding critical exponent. We may imagine a nonequilibrium state (e.g., a fluctuation) that relaxes to the equilibrium state with its characteristic (relaxation) time close to the phase transition scaling as $\tau_c \sim \xi^{-z} = t^{-\nu z}$, where z is called the *dynamical critical exponent*. Note that we have assumed implicitly that the dynamic properties near the critical point also scale with ξ. The most important situation arises when this relaxation time is related to the characteristic time for quantum processes via the uncertainty principle $\omega_c \cdot \tau_c \simeq 1$, where the characteristic energy for those processes is $\hbar\omega_c \gtrsim k_B T_c$, so that $k_B T_c/(\hbar \cdot \tau_c) \lesssim 1$, that is,

$$T_c \lesssim \frac{\hbar}{k_B} \cdot \frac{1}{\tau_c} \sim t^{\nu z}. \tag{16.1}$$

A Primer to the Theory of Critical Phenomena. http://dx.doi.org/10.1016/B978-0-12-804685-2.00022-X

In that situation the value z is also crucial in determining T_c and other equilibrium critical properties of the system. This very fact introduces a new aspect when approaching T_c. The principal question at this point is the physical nature of the characteristic energy $\hbar\omega_c$ of the (quantum) fluctuations, with τ_c representing the relaxation time. In that situation, $T \to 0$ also means that $T_c \to 0$, that is, we are indeed entering the true quantum regime. *The situation is quite singular.* Namely, we enter into the regime $T \to 0$, accompanied by a singular behavior of the physical properties. So, the naive concept that the statistical-mechanical equilibrium state reduces to a pure quantum-mechanical state as $T \to 0$ does not apply to this case: physical quantities such as C_p and χ, which can be defined for any temperature, may now approach singular values at very low T. This also means that peculiar conditions have to be met for the system to exhibit *quantum critical behavior.* For example, concepts such as a finite Pauli magnetic susceptibility χ, or Landau effective mass m^*, or other well-defined quantities for a metal, must all be abandoned, or at least be radically modified. In other words, we enter the world of strange metals, electron localization, or magnetic instabilities at $T = 0$.

We may ask what is the principal factor determining the critical point at a continuous phase transition as $T_c \to 0$. It cannot be the statistical entropy because, as $T \to 0$, this is then a continuous function, since $S = -(\partial F/\partial T)_V$. A direct answer to this question is the competing (mutually compensating) interactions between the particles (or spins) composing the system. Formally, we may have two terms in the Hamiltonian, which do not commute with each other; then a singular situation arises when one of those terms has either soft (zero-energy) modes or low-energy modes with $\omega \simeq \omega_c \sim k_B T$ at any nonzero T.

In Fig. 16.1, we show a schematic phase diagram for the temperature–pressure plane, which also contains the quantum critical point (QCP) evolving at the critical temperature T_c. The regime in red marks the classical-critical behavior regime that reduces gradually to a point as $T_c \to 0$. In this figure, we have depicted a system of itinerant electrons that under pressure undergoes a continuous ferro- to paramagnetic transition, that is, from a ferromagnetic metal to a normal metal, called a Fermi liquid. The most interesting regime is the region between those two states; this is called *a non-Fermi (non-Landau)* quantum-liquid regime. In this regime, as $T \to 0$ (as marked by the vertical arrow), new scaling laws appear for physical quantities, such as the magnetic susceptibility or specific heat, that are entirely different from those in the classical regime. Some of these laws are briefly discussed later in this chapter.

We first introduce a simple example of a localization–delocalization transition for electrons as an extreme, but still physically clear, illustration of quantum-critical behavior, before turning to a more formal approach.

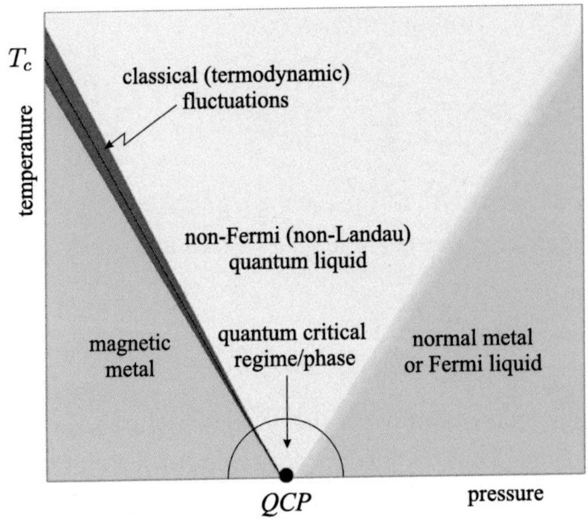

FIGURE 16.1 Representation of physical regimes in the neighborhood of the quantum critical point (QCP). This arises in a metallic system with correlated electrons. The phase within the circle enclosing the QCP may be that of a superconductor (superfluid). A nonanalytic singular behavior is observed in the encircled regime when $T \to 0$, that is, as the system gradually reaches the pure quantum limit (thermodynamic fluctuations, marked in red and discussed so far, play only a partial role and vanish at $T = 0$). For details, see the main text.

16.2 EXAMPLE: THE MOTT–WIGNER CRITERION FOR THE ELECTRON GAS INSTABILITY

We have selected a system of itinerant electrons, that is, the case of an interacting electron gas, because most of the quantum-divergent behavior is observed for strongly correlated electronic systems.

In an ideal, that is, noninteracting electron gas, the statistical distribution of particle energies is provided by the Fermi–Dirac distribution function[1]

$$n_{\boldsymbol{k}\sigma} = \frac{1}{\exp\left[\beta\left(\epsilon_{\boldsymbol{k}\sigma} - \mu\right)\right] + 1}, \tag{16.2}$$

where $n_{\boldsymbol{k}\sigma}$ is the average number of electrons with momentum $\hbar\boldsymbol{k}$, and the spin quantum number is $\sigma = \pm 1$ (note that its physical spin z-component is $s^z = (\hbar/2)\sigma$), $\epsilon_{\boldsymbol{k}\sigma}$ is the particle energy in that state, and μ is the chemical potential, determined from the condition that the total number of particles N

1. For an elementary discussion of the ideal Fermi gas, see, for example, L.D. Landau and E.M. Lifshitz, *Statistical Physics* (Elsevier, Amsterdam, 2005), Chapter V; K. Huang, *Introduction to Statistical Physics* (Taylor & Francis, London, 2001), Chapter 9.

should be conserved; that is, from the equation[2]

$$N = \sum_{k\sigma} n_{k\sigma} = 2\sum_{k} \frac{1}{\exp\left[\beta\left(\epsilon_{k} - \mu\right) + 1\right]}. \tag{16.3}$$

As $T \to 0$ the canonical properties of the electron gas are recovered. For example, the average kinetic energy per particle at $T = 0$ in this ideal 3d gas is

$$\bar{\epsilon} \equiv \frac{3}{5}\frac{\hbar^2}{2m}\left(3\pi^2\frac{N}{V}\right)^{2/3}, \tag{16.4}$$

where m is the particle mass, and V is the system volume. In deriving this expression, one utilizes the Pauli exclusion principle for fermions. If we define the particle density as $\rho \equiv N/V$, then for a 3d gas the energy increases with increasing density as $\sim \rho^{2/3}$. On the other hand, the average Coulomb repulsive interaction per particle is

$$\epsilon_{e-e} = \frac{e^2}{2\varepsilon r_s} = \frac{e^2}{2\varepsilon}\left(\frac{N}{V}\right)^{1/3} = \frac{e^2}{2\varepsilon}\rho^{1/3}, \tag{16.5}$$

where ε is the static dielectric constant of the system, and V/N is the effective volume per particle, so that the average classical distance between the particles is $r_s = (V/N)^{1/3}$. We see that, as the particle density increases, the kinetic energy begins to increase faster ($\sim \rho^{2/3}$) than the interaction energy, which is proportional to $\sim \rho^{1/3}$, once we exceed a certain value of ρ. One can ask what happens when those two energies coincide, that is, when $\bar{\epsilon} = \epsilon_{e-e}$. An elementary inspection of this equality leads to the condition

$$a_B\, \rho_c^{1/3} \simeq 0.2, \tag{16.6}$$

where $a_B \equiv \hbar^2/(me^2)\varepsilon$ is the effective Bohr radius for an electron in a medium with dielectric constant ε, and ρ_c, the density at this critical point. This condition is called *the Mott* (or *Mott–Wigner*) estimate for the electron localization threshold of the particles. Note that $\rho_c^{1/3} = r_c^{-1}$, so the effective distance between the "lattice" of the frozen (localized) particles is then of the order of $5a_B$.

The basis for of this reasoning is as follows: For $\rho < \rho_c$ the interaction energy $\epsilon_{e-e} > \bar{\epsilon}$, whereas for high densities ($\rho > \rho_c$), the kinetic energy dominates. At the extreme densities $\rho \gg \rho_c$, $(\epsilon_{e-e}/\bar{\epsilon}) \to 0$; hence, the interacting gas approaches the limit of an ideal (noninteracting) gas, even though $\epsilon_{e-e} \to \infty$.

2. This is a nontrivial condition, as the derivation of formula (16.2) is based on the assumption that we can have an arbitrary occupation of each energy level. This is not the case because we deal with a finite number of particles.

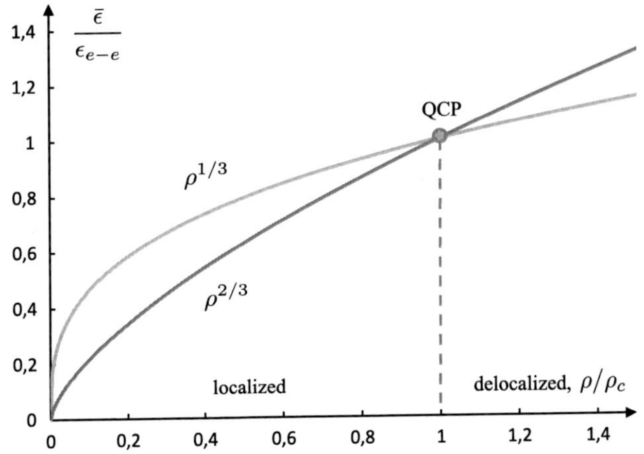

FIGURE 16.2 Schematic illustration of the transition between the delocalized (plane-wave) states of an ideal electron gas and the localized (atomic, frozen) configurations of the Mott–Wigner type. This transition requires the further analysis carried out below. The discussion relies on emphasizing the competition between kinetic energy and repulsive interparticle Coulomb interactions.

Closer inspection also leads to the conclusion that for $\rho < \rho_c$, the states are localized, since then the repulsive interaction forces dominate; i.e., the particles spread as far away from each other as possible (to the distance r_c). Parenthetically, this is also the reason why the effective Bohr radius appears out of nowhere at this "*critical point*" ρ_c, even though there is no sign of any attractive atomic centers on which the electrons can form atomic states. The mutual energetics are illustrated in Fig. 16.2.

All this elementary reasoning shows that the competing energies suffice to lead to a qualitative change of character of physical states. Next, we show on a slightly modified model, the Hubbard model, that this peculiar physical behavior can indeed lead to a true phase transitions in the $T \to 0$ limit for realistic (lattice) systems, starting with well-defined atomic states at large interatomic distances.

16.3 LOCALIZATION ON A LATTICE: THE HUBBARD MODEL

We start with a lattice of atoms, each atom contributing one electron to the system. Effectively, we can imagine a lattice of hydrogenic-like atoms, with one valence electron, each in the 1s-like state. Only the lattice of those 1s states concerns us. The question is rather, under what conditions the assembly of electrons sits on atoms (in localized atomic states) and when, instead, they form a collective gas of interacting electrons, confined by a skeleton of positive ions. The latter play only a passive role in the whole process. The situation is

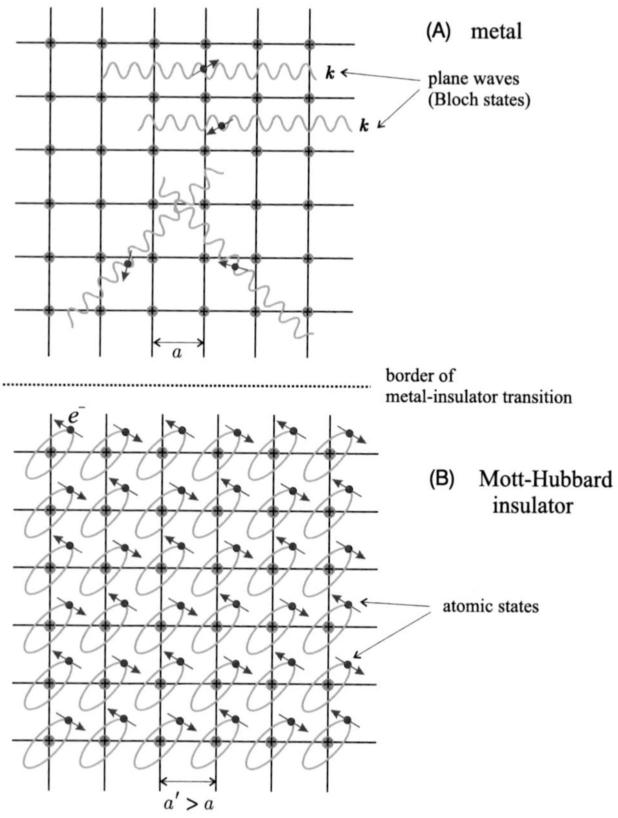

FIGURE 16.3 Illustrative representation of the metallic (A) and Mott insulating (B) (localized) states. In state (A) the electrons are derived from the parent atoms (only positive cations K^+ are left) and still form a lattice, as even in the delocalized state the electrons are regarded as the only relevant quantum particles.

schematically depicted in Fig. 16.3. The question is: what happens in situations intermediate between the two pictures?

The simplest Hamiltonian describing such an interacting gas on the lattice was proposed by Hubbard (1963); its explicit form is

$$\mathcal{H} = \sum_{\boldsymbol{k}\sigma} \epsilon_{\boldsymbol{k}\sigma}\, n_{\boldsymbol{k}\sigma} + U \sum_{i} n_{i\uparrow}\, n_{i\downarrow}. \tag{16.7}$$

Here $\epsilon_{\boldsymbol{k}}$ is the energy of an individual particle in the spin nonpolarized state with the wave vector $\boldsymbol{k}$, $n_{\boldsymbol{k}\sigma}$ is, as earlier, the number of particles in the single-particle state characterized by the quantum numbers $(\boldsymbol{k}, \sigma)$, $n_{i\sigma} = 0, 1$ is the number of electrons on a given lattice site i with spin σ, and U is the magnitude of the contact interaction when two particles with opposite spins meet on the same atomic site i. At first glance, this Hamiltonian has no connection to the reasoning of the preceding section, but this is not quite the case. The similarity is the following: The energy $\epsilon_{\boldsymbol{k}}$ varies in the interval $[-W/2, W/2]$ (we assume that the reference atomic energy of each particle can be taken to be $\epsilon_{at} = 0$). The quantity W is called the bandwidth; the single-particle part (the first term) is more negative (the energy is smaller) if W increases. If we assume that the energies are spread uniformly, then the average energy per particle is $-W/4$. The interaction energy per particle is $U\langle n_\uparrow n_\downarrow\rangle$. For one particle per atom, the mean-field estimate is $\langle n_{i\uparrow} n_{i\downarrow}\rangle \simeq \langle n_i \uparrow\rangle\langle n_i \downarrow\rangle = 1/4$. Effectively, the two energies balance each other when they are of the same magnitude,

$$F = -W/4 + U/4 = 0, \tag{16.8}$$

that is, $U = W$. This condition replaces the former condition (16.6) and physically amounts to the same thing. Namely, if $W > U$, the single-particle ("kinetic") energy dominates, and the particles form a fluid with $\{\boldsymbol{k}\}$ as a good quantum number. In the opposite limit, the double occupancies, characterized by joint probability $\langle n_{i\uparrow} n_{i\downarrow}\rangle$, vanish. All this requires a more careful study, as the last statements involve assumptions basic to the more detailed quantitative approach provided below.

16.4 DETAILED DISCUSSION OF LOCALIZATION OF FERMIONS: QUANTUM CRITICAL POINTS

The basic question we face when discussing, for example, electronic systems with a localization–delocalization transition is as follows: The kinetic (now band) energy of electrons in the delocalized state is of the order $W \sim$ 1–3 eV per particle; hence, $U \gtrsim W$. The thermal energy in the range of interest to us is $k_B T \simeq 10$ meV ~ 100 K. The principal question is then how it is possible that such a feeble external thermal stimulus can drastically change the character of electronic states from itinerant to localized states or vice versa. Note that this is a macroscopic state involving of the order of 10^{23} particles. The simplest answer is that the energies in (16.8) that are close to the transition almost compensate each other, thus making the system extremely sensitive to feeble perturbations such as the thermal stimulus, external pressure, or even chemical doping. Parenthetically, the metallic state is universally regarded as a robust state of matter. Another class is the state with localized electrons, regarded throughout the book, as another universal state. The question is: when do they become unstable relative to each other?

To put this qualitative argument on a quantitative basis, we assume that the correlation function $\eta \equiv \langle n_{i\uparrow} n_{i\downarrow} \rangle$ involved in Eq. (16.7) is a new basic physical parameter, in addition to the individual particle energies $\{\epsilon_{\boldsymbol{k}}\}$ and to the interaction parameter U. The parameter η differentiates between the delocalized and localized states in the following manner:

$$\eta = \begin{cases} \langle n_{i\uparrow} \rangle \langle n_{i\downarrow} \rangle \simeq \frac{1}{4} \text{ for } U = 0; \\ 0 \text{ for } U \to \infty. \end{cases} \tag{16.9}$$

We see that in the simplest case, η plays the role of an order parameter. The task is to determine the explicit form of the thermodynamic Landau-type function of the parameter η.

To accomplish the above two tasks, we make a further assumption that since the two terms composing the total energy (16.7) are of comparable magnitude, the interaction of the kinetic energy of the initially independent particles must be renormalized in the following manner:

$$\epsilon_{\boldsymbol{k}} \to \Phi \epsilon_{\boldsymbol{k}} \equiv E_{\boldsymbol{k}}, \quad 0 \leqslant \Phi \leqslant 1, \tag{16.10}$$

where $\Phi \equiv \Phi(\eta)$ with the following properties:

$$\Phi(\eta) = \begin{cases} 1 \text{ for } \eta = \langle n_{i\uparrow} \rangle \langle n_{i\downarrow} \rangle = \frac{1}{4} \ (U = 0), \\ 0 \text{ for } \eta = 0 \ (U \geqslant U_c). \end{cases} \tag{16.11}$$

These two conditions show respectively that the band energy takes the usual form ($E_{\boldsymbol{k}} = \epsilon_{\boldsymbol{k}}$) when $U \ll W$, and that the band energy and the interaction energies disappear when $U \gg W$ ($\Phi \epsilon_{\boldsymbol{k}} = U\eta = 0$). This last condition amounts to what we term as a perfect compensation of the two competing energies.

In the complete formulation, we must also deal the situation where $U \to W = 0$. Then, in the spirit of the Landau expansion, we set $\Phi(\eta) = f_0 + f_1\eta + f_2\eta^2 + \cdots$. The coefficients f_0, f_1, and f_2 are determined from the known limiting situations (16.11) and from the fact that $\eta = 1/4$ for $U = 0$. The two conditions modify expression (16.7), which now reads

$$E \equiv \langle \mathcal{H} \rangle / N = \Phi(\eta)\bar{\epsilon} + U\eta. \tag{16.12}$$

In addition, the explicit expression for η is obtained from the Landau-type minimization condition $\partial E / \partial \eta = 0$ for $\eta = 1/4$, which leads to the explicit result for $\Phi(\eta)$, namely

$$\Phi(\eta) = 8\eta(1 - 2\eta). \tag{16.13}$$

After minimization of E with respect to η all involved quantities read:

$$\eta \equiv \eta_0 = (1/4)(1 - U/U_c), \tag{16.14}$$

$$\Phi \equiv \Phi_0 = \left[1 - \left(\frac{U}{U_c}\right)^2\right], \tag{16.15}$$

$$E_G = \bar{\epsilon}\left(1 - \frac{U}{U_c}\right)^2, \tag{16.16}$$

where now $U_c \equiv 2W$ and $\bar{\epsilon} = -W/4$. We see that those quantities scale with U/U_c; hence, U_c plays the role of a critical value, up to which the metallic (delocalized) state of particles is stable. In the present formulation, we cannot go beyond the value $U = U_c$. Namely, for $U > U_c$, we encounter atomic states on lattice, that is, *the Mott–Hubbard insulating state*. For all practical purposes this is *the Heisenberg-antiferromagnet state* depicted in Fig. 16.3B. However, the inclusion of an antiferromagnetic state requires an essential extension of the present approach. We need to incorporate, among other items, exchange interactions in the insulating phase. This in turn is the starting point for the discussion of the low-T regime where the metal–insulator transition also involves the magnetic states. This last subject is not discussed here. Instead, we quote the results,[3] which document the critical behavior as $U \to U_c \equiv 8|\bar{\epsilon}| = 2W$. Namely, the static magnetic susceptibility at $T = 0$ of the system in the delocalized state is

$$\chi = \chi_P/\Phi_0\left[1 - \frac{U}{W}\frac{1 + U/(2U_c)}{(1 + U/U_c)^2}\right], \tag{16.17}$$

where χ_P is the Pauli susceptibility for the noninteracting electron gas. If $U \to U_c - 0$, then[4]

$$\chi \sim \frac{\chi_P}{\Phi_0} \to \infty \quad \text{as} \quad U \to U_c. \tag{16.18}$$

This is a clear sign of a singularity if the phase transition is continuous. Probably the most spectacular feature of this transition is the magnitude of the spin $\boldsymbol{S}_i$ as $U \to U_c$. It is given by

$$\langle \boldsymbol{S}_i^2 \rangle = \frac{3}{8}\left(1 + \frac{U}{U_c}\right) \longrightarrow \frac{3}{4} \quad \text{as} \quad U \to U_c. \tag{16.19}$$

3. For a detailed summary of the results obtained for such a mean-field picture, see, e.g., J. Spałek, *J. Sol. St. Chem.* **88**, 77–93 (1990).

4. Note also that another singularity arises when the other factor $[\ldots] = 0$. This corresponds to the so-called Stoner singularity for the onset of ferromagnetism, which is not studied here.

Note that, as $U \to U_c$, $\langle \boldsymbol{S}_i^2 \rangle \to \frac{3}{4} \equiv \frac{1}{2}\left(\frac{1}{2}+1\right)$, that is, we recover the full atomic spin, as has been assumed throughout the book for a Heisenberg magnet with spin $S = 1/2$. This result shows clearly under which conditions we can regard the system as a lattice composed of $(1/2)$ spins, as assumed explicitly throughout the book. This was not at all a trivial assumption, as it requires the formation of stable atomic states in a periodic lattice. Simply stated, the electrons in atomic states in a solid persist as occupied states, since the energy of motion (hopping) from site to site by individual particles would require an increase of energy by U, which exceeds the gain in kinetic energy $(-W/2)$. *This is the reason for the existence of atomic spins in a solid.*

16.5 QUASIPARTICLE REPRESENTATION OF INTERACTING ELECTRON SYSTEMS: FROM CLASSICAL TO QUANTUM CRITICAL POINTS

So far we have dealt with the $T = 0$ transformation of itinerant to localized states or vice versa. The fact that the physically measurable quantities diverge tell us that, at least within this simple scheme, the many-particle quantum states can then become singular. Obviously, we are not able to do measurements at $T = 0$ directly. Therefore, a natural question arises whether we can generalize the above results to the $T > 0$ regime, which would reduce to those just discussed in the limit $T \to 0$. This is the question we are going to tackle next.

First, we should note that the single-particle part of energy in (16.12) is multiplied by $\Phi(\eta)$. Here we introduce, what may seem natural, renormalized individual single-particle states, that is, we assume that $E_{\boldsymbol{k}} = \Phi\epsilon_{\boldsymbol{k}}$. The reason for this renormalization of individual quantum single-particle states is that at $T = 0$ we can write

$$\Phi(\eta)\bar{\epsilon} = \frac{1}{N}\Phi\sum_{\boldsymbol{k}\sigma}\epsilon_{\boldsymbol{k}}\, n_{\boldsymbol{k}\sigma} \equiv \frac{1}{N}\Phi\sum_{\substack{k<k_F\\ \sigma}}\epsilon_{\boldsymbol{k}} = \frac{1}{N}\sum_{\substack{k<k_F\\ \sigma}}(\Phi\epsilon_{\boldsymbol{k}}) = \sum_{\substack{k<k_F\\ \sigma}} E_{\boldsymbol{k}}. \tag{16.20}$$

Hence, in this (correlated) quantum liquid the correlated character is acquired by the individual energies $E_{\boldsymbol{k}} = \Phi\epsilon_{\boldsymbol{k}}$. We also assume that in first approximation quasiparticles of energy $E_{\boldsymbol{k}}$ still obey the Fermi–Dirac distribution[5] at $T > 0$, that is,

5. This assumption is in the spirit of the Landau (1956) theory of Fermi liquids; cf. J. Spałek, *Reference Module in Materials Science and Materials Engineering* (Elsevier, Oxford, 2016), pp. 1–20. However, we must emphasize that the Landau theory of Fermi liquids does not involve any discussion of quasiparticle localization into the Mott state of localized spins.

$$n_{\mathbf{k}\sigma} \equiv f_{\mathbf{k}\sigma} = \frac{1}{\exp\left[\beta\left(E_{\mathbf{k}\sigma} - \mu\right)\right] + 1}, \tag{16.21}$$

where μ is the chemical potential. For the case of one electron per atom, we can take $\mu \equiv 0$, that is, in the middle of the band energy $\epsilon_{\mathbf{k}}$, since we assume that particle–hole symmetry holds; that is, the energy is symmetric with respect to the middle point of band. Note that we have put the quasiparticle energies into the Fermi function, in the spirit of the Landau theory.

With those assumptions, we can explicitly write down the Landau-type free-energy functional for the system of fermions in the usual manner, namely[6,7]

$$\frac{\mathcal{F}}{N} = \frac{1}{N}\sum_{\mathbf{k}\sigma} E_{\mathbf{k}} f_{\mathbf{k}\sigma} + U\eta + \frac{k_B T}{N}\sum_{\mathbf{k}\sigma}\left[f_{\mathbf{k}\sigma}\ln f_{\mathbf{k}\sigma} + (1 - f_{\mathbf{k}\sigma})\ln(1 - f_{\mathbf{k}\sigma})\right], \tag{16.22}$$

where $\eta = \eta(T)$ plays, as before, the role of the order parameter. The first two terms represent the effective internal energy, whereas the last is the entropy contribution in a given configuration characterized by $\eta(T)$. The true free energy (equal to the Gibbs energy as $\mu \equiv 0$) is found from the minimum conditions:

$$\frac{\partial \mathcal{F}}{\partial \eta} = 0; \quad \frac{\partial^2 \mathcal{F}}{\partial \eta^2} > 0. \tag{16.23}$$

The value of $\eta \equiv \eta(T)$ obtained in this manner, when substituted into Eq. (16.22), determines the physical free energy $F(T)$. We see that the simplest (mean-field-type) theory is much more involved for quantum particles (fermions) as compared to those for pure spin systems, as this theory contains a logarithmic contribution to the entropy that cannot be expanded in powers of the order parameter, as was the case in the earlier chapters. Note also that, in the $T = 0$ limit, the results reduce to Eqs. (16.14)–(16.16), as should be the case.

The analysis of solution of (16.22) is quite cumbersome and in the general case must be carried out numerically.[6] For the purpose of discussing the classical and quantum critical points, it is sufficient to limit ourselves to the low-temperature limit, which is overviewed next. The result for the Landau function expression to order T^2 is[7]

$$\frac{\mathcal{F}}{N} = -\Phi\frac{W}{4} + U\eta - \frac{\gamma_0 T^2}{\Phi} + o(T^4). \tag{16.24}$$

6. J. Spałek, A. Datta, and J.M. Honig, *Phys. Rev. B* **33**, 4891 (1986); J. Spałek, *J. Sol. St. Chem.* **88**, 70–93 (1990).
7. J. Spałek, A. Datta, and J.M. Honig, *Phys. Rev. Lett.* **59**, 728 (1987); J. Spałek, M. Kokowski, and J.M. Honig, *Phys. Rev. B* **39**, 4175 (1989).

The first two terms appeared before; the third represents the thermal excitations across the Fermi level located at $\mu \equiv 0$. The quantity γ_0/Φ represents the renormalized linear-specific heat (Sommerfeld) coefficient with

$$\gamma_0 = \frac{2\pi^2}{3} k_B^2 \rho(0) \tag{16.25}$$

being the unrenormalized form, where $\rho(0)$ is the density of unrenormalized states at the Fermi level (per atom per spin). In our simplified discussion, $\rho(0) = 1/W$. One additional warning is in place. Namely, Eq. (16.23) becomes singular as $\Phi \to 0$ with $T > 0$. Therefore, a singular jump to the state $\eta = 0$ is possible. Fortunately, physics comes to the rescue to deal properly with singularity. Namely, as we have noted earlier, the fermionic liquid transforms to a system of localized moments. In the latter state, the free energy is also nonzero. Neglecting the exchange interaction among the spins (magnetic order ignored), it can be written in the form

$$\frac{F_{ins}}{N} = -k_B T \ln 2, \tag{16.26}$$

that is, we include only the entropy part of freely fluctuating spins with $S = 1/2$.

In effect, if we are interested in a discontinuous (first-order) transition, the required condition is $F_{met} = F_{ins}$, where F_{met} is the physical free energy of the delocalized (metallic) state, obtained after minimization of $\mathcal{F}$ with respect to η, combined with its low-T expansion. In what follows, we discuss only the final results.[5,6,7]

Let us define $I \equiv U/U_c$. The functional (16.24) transforms into an intuitively clear expression for the free energy:

$$\frac{F_{met}}{N} = (1 - I)^2 \bar{\epsilon} - \frac{\gamma_0 T^2}{1 - I^2} + o(T^4). \tag{16.27}$$

Equating Eqs. (16.26) and (16.27), we obtain two transition temperatures. Explicitly,

$$\frac{k_B T_\pm}{W} = \frac{3}{2\pi^2}\left[1 - \left(\frac{U}{U_c}\right)^2\right]\left\{(\ln 2)^2 \pm \left[(\ln 2)^2 - \frac{1 - U/U_c}{1 + U/U_c}\right]^{1/2}\right\}. \tag{16.28}$$

This result is depicted in Fig. 16.4 (as before, $U_c = 2W$). The lower temperature branch (T_-) is drawn as a solid line, whereas the T_+ part is actually a crossover line, not properly accounted for in our low-T expansion. The T_- curve ends at a classical critical point, which determines the lowest value of $U = U_{lc}$,

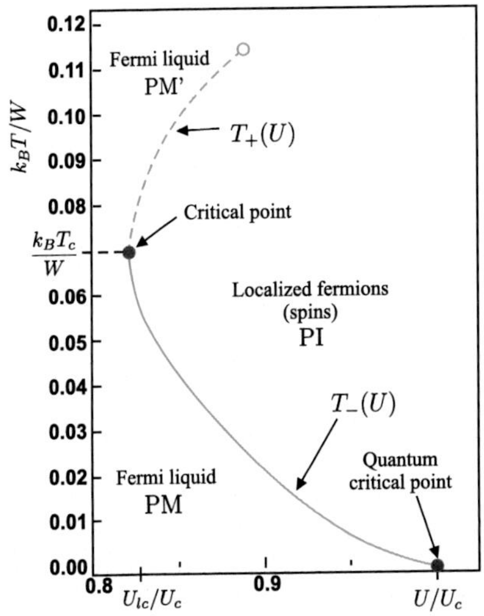

FIGURE 16.4 Phase diagram in the paramagnetic (PM) region of the metal-insulator transition. The critical points and characteristic temperatures are marked explicitly; the points (T_c, $T_c = 0$) are at the end points of the first-order (blue solid) line.

via the condition $T_+ = T_-$, below which the delocalized states are always stable. The value of the critical temperature T_c is

$$\frac{k_B T_c}{W} = 1 - \frac{3\ln 2}{2\pi^2}\left[1 - \left(\frac{U_{lc}}{U}\right)^2\right]. \tag{16.29}$$

The value of $T_c = 0$ appears for $U = U_c = 2W$. This critical point can be called a *quantum critical point* (QCP), as the entropy is zero and the two mechanical energies compensate each other. On the other hand, the solid line between T_c and zero represents a line of discontinuous transitions, as is illustrated further in Fig. 16.5, where we have drawn the free energies of the two phases. The curves $a - d$ depict the evolution of the Fermi liquid state in the low $U < U_{lc}$ region at the limit as $U \to U_c$. When the parabola is tangent to the straight line e (between a and b), we reach the lower critical value U_{lc} for the transition to take place. Furthermore, for U between U_{lc} and U_c, we observe two transitions. The point $U = U_c$ is singular, as both curves coalesce into a single QCP. Moreover, for $U \in (U_{lc}, U_c)$, the two curves intersect at a nonzero angle; then the entropy $S = -\partial F/\partial T$ is discontinuous at the transition. Hence, it is a first-order transition, except at the points $U = U_{lc}$ and U_c. Strictly speaking, the QCP becomes a *hidden critical point* when we include the exchange interaction in the localized phase,[6] but this topic will not be touched upon here.

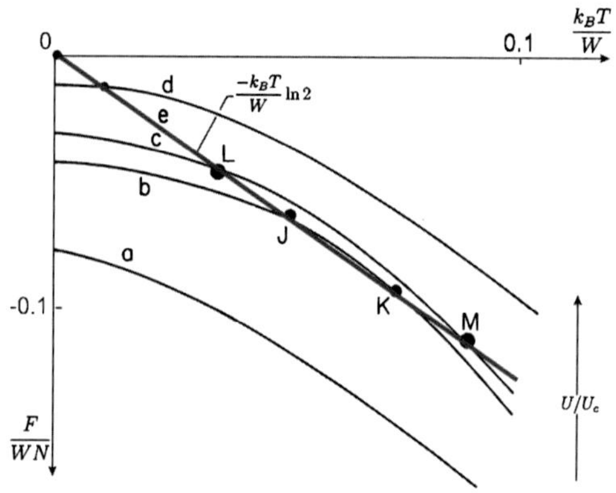

FIGURE 16.5 Plots of temperature dependence of the free energy in the paramagnetic state: The parabolas $a - d$ represent the energy of the metallic (Fermi-liquid) state with ascending U/U_c ratio. The straight line e (in red) represents the energy of the Mott insulating (PI) state. The crossings L and J correspond to a discontinuous PM $\to$ PI transition, whereas those at K and M correspond to the reverse PI $\to$ PM' transition. The transition to the high-T PM metallic phase can involve crossover behavior. Note the competition between the energy and the entropy contribution in creating the two transitions, both of the Mott type at $T > 0$.

We should point out that the cursory discussion presented here has its full experimental confirmation in the observed transitions for the compounds V_2O_3 and $NiS_{2-x}Se_x$.

16.6 QUANTUM PHASE TRANSITION: A GENERALIZED LANDAU–HERTZ FUNCTIONAL

As we have seen, the conditions of equilibrium paramagnetism near the continuous phase transitions for the critical temperature $T_c > 0$ (see, e.g., Chapter 13) did not require the inclusion of time dynamics of the order parameter fluctuations. This is not the case when $T_c \to 0$, as we have mentioned in Section 16.1. The time (or equivalently frequency) domain is best included by regarding the inverse temperature as an imaginary time $\tau = -i\hbar\beta$ (the procedure known as the Wick rotation), so that the partition function $\mathcal{Z} = \mathrm{Tr}(e^{-\beta\hat{H}})$ can be regarded as a time-evolution operator $e^{-i\hat{H}\tau/\hbar}$. In this language the partition function, with inclusion of fluctuations, can be written in the form

$$\mathcal{Z} = \int \mathcal{D}[\bar{\Psi}, \Psi]\, e^{S(\bar{\Psi}, \Psi)}, \tag{16.30}$$

where the fields $\bar{\Psi}$ and Ψ are generalization of the concept of the local order parameter, $\Phi(\boldsymbol{r}) \to \Psi(\boldsymbol{r}, \tau)$, $\Phi^*(r) \to \bar{\Psi}(\boldsymbol{r}, \tau)$, and these new fields are regarded as *quantum fields*. In that language, the operator S is called the *action*; and its explicit form is

$$S(\bar{\Psi}, \Psi) = \int d^d\boldsymbol{x} \int_0^\beta d\tau \left\{ \bar{\Psi}(\boldsymbol{x}, \tau) \left[-\frac{\partial}{\partial_\tau} + \mu \right] \Psi(\boldsymbol{x}, \tau) \right. \\ \left. - H\left(\bar{\Psi}(\boldsymbol{x}, \tau), \Psi(\boldsymbol{x}, \tau) \right) \right\}, \tag{16.31}$$

where now the Hamiltonian is expressed in terms of the new fields. When comparing the last expression with Eq. (11.6), we see that the integration in the exponential part of $\mathcal{Z}$ now looks like that of a $(d+1)$-dimensional Euclidean space–time integral, where the additional temporal dimension is limited to the interval $[0, \beta]$. As $T \to 0$, we see the infinite upper limit for β as that for an effective classical system. The quantity μ is the chemical potential, since the number of particles should be conserved.

Formally, the new (first term) is the presence of the time derivative (∂/∂_τ). The simplest model we can imagine is replacing the remaining terms by the Landau functional. In other words, we can postulate the simplest action in the form

$$S(\bar{\Psi}, \Psi) = \int d^d\boldsymbol{x} \int_0^\beta d\tau \left[\frac{a}{2} \left(\partial_\tau \bar{\Psi}\right) \left(\partial_\tau \Psi\right) + \frac{c}{2} \left(\nabla \bar{\Psi}\right) \cdot \left(\nabla \Psi\right) \right. \\ \left. + \frac{r}{2} \bar{\Psi}\Psi + \frac{v}{4} \left(\bar{\Psi}\Psi\right)^2 \right], \tag{16.32}$$

where now $|\Phi|^2 \to \bar{\Psi}\Psi$, etc. This "simple" addition leads to far-reaching consequences in the formal solution and calculation of physical properties as $T_c \to 0$. Two methodological remarks are in place. First, we formulate the theory for $T > 0$, and then discuss the $T \to 0$ limit. In that way, we can make experimentally verifiable predictions concerning the scaling in the low-temperature regime. Second, the terms with derivatives are both positive; this is the reason why we regard the integral (16.32) as operating in Euclidean space.

Formal Remark

The functional (16.32) can be brought into a format where all the coefficients a, c, r, and v equal to unity. This procedure is carried out by rescaling the first term $\tau \to \tau/a$, $\bar{\Psi} \to \bar{\Psi}_0$, and $\Psi \to \Psi_0$, where Ψ_0 is the mean-field homogeneous solution ($\Psi_0 = (r/v)^{1/2}$), and the spatial coordinates $x_i \to x_i/\xi$, where ξ is the mean-field (Ornstein–Zernike) correlation length ($\xi = (c/r)^{1/2}$). In such a rescaling τ is expressed in units of a characteristic frequency $(1/a)$, spatial coordinates in units of the fluctuation size, while the order parameter is expressed in units of its homogeneous static value.

16.7 AN EXAMPLE

The order parameter depends on both $\boldsymbol{x}$ and τ; the latter variable varies in a finite interval $[0, \beta\hbar]$ for $\beta < \infty$. It turns out convenient to transform Eq. (16.32) to the frequency (ω) and momentum $(\boldsymbol{k})$ domains. However, the first transform is carried out in a finite interval $\beta\hbar$, which provides the periodicity for Ψ. Explicitly,

$$\Psi(\boldsymbol{x}, \tau) = \frac{1}{\beta} \int \frac{d^d k}{(2\pi)^d} \sum_{\omega_n} \Psi(\boldsymbol{k}, \omega_n)\, e^{i(\boldsymbol{k}\cdot\boldsymbol{x} + \omega_n \tau)} \tag{16.33}$$

with the so-called *Matsubara frequencies*

$$\omega_n = \begin{cases} 2n\pi k_B T & \text{for Bose systems,} \\ (2n+1)\pi k_B T & \text{for Fermi systems,} \end{cases} \tag{16.34}$$

for $n = 0, \pm 1, \pm 2, \ldots, \pm\infty$. We see that in quantum phase transitions the field Ψ depends on the multiparticle system statistics[8] it describes.

In the interesting case of the extended Landau functional, Eq. (16.32), the frequencies $\{\omega_n\}$ are selected as for Bose systems. We quote here only the result for the paramagnetic to ferromagnetic transition as described by the Hubbard model at a microscopic level. The explicit form of the functional (16.32) in the $(\boldsymbol{k}\omega)$ representation then is[9]

$$S \equiv S_{eff}[\Psi] = S^{(0)} + S^{(2)}_{eff}[\Psi] + S^{(4)}_{eff}[\Psi] + \cdots \tag{16.35}$$

with

$$S^{(2)}_{eff} = \frac{1}{2}\beta V \int \frac{d^d k}{(2\pi)^d} \sum_{n=-\infty}^{\infty} \left[\delta + k^2 + \frac{|\omega_n|}{|\boldsymbol{k}|}\right] \Psi_n(\boldsymbol{k})\, \Psi_{-n}(-\boldsymbol{k}) \tag{16.36}$$

now representing the Gaussian part and

8. For a more accurate discussion, see, e.g., P. Coleman, *Introduction to Many-Body Physics* (Cambridge University Press, 2015), Ch. 8.
9. J.A. Hertz, *Phys. Rev. B* **14**, 1165 (1976); A.J. Millis, *ibid.*, **48**, 7183 (1993).

TABLE 16.1 Temperature dependences of basic physical quantities listed in the vicinity of QCP for ferromagnetic (FM) and antiferromagnetic (AFM) metals of d spatial dimensions[11]; z is the dynamical critical exponent, and $\Delta\rho$ and $\Delta\chi$ denote the corresponding contributions appearing in the critical regime. In the logarithmic functions, $\ln T$ should always be understood as $\ln(T/T_0)$, where $k_B T_0$ is the characteristic energy ($\hbar\omega_c$) of dynamical excitations; here they are the spin fluctuations in the single-band Hubbard model

Dimension	Transition type	Specific heat, C/T	Resistivity, $\Delta\rho$	Magnetic susceptibility, $\Delta\chi$
$d=3$	FM ($z=3$)	$\sim \log T_0/T$	$\sim T^{5/3}$	$\sim T^{-4/3}$
	AFM ($z=2$)	$\sim 1-aT^{1/2}$	$\sim T^{3/2}$	$\sim T^{-3/2}$
$d=2$	FM ($z=3$)	$\sim T^{5/3}$	$\sim T^{3/2}$	$\sim T^{-1}/\log T$
	AFM ($z=2$)	$\sim \log T_0/T$	$\sim T$	$\sim (\log T)/T$

$$S_{eff}^{(4)} = \frac{v}{4}\beta V^4 \sum_{n_1 \ldots n_4} \left(\prod_{i=1}^{4} \int \frac{d^d k_i}{(2\pi)^d}\right) \delta\left(\boldsymbol{k}_1 + \boldsymbol{k}_2 + \boldsymbol{k}_3 + \boldsymbol{k}_4\right) \delta_{n_1+n_2+n_3+n_4}$$
$$\times\, \Psi_{n1}(\boldsymbol{k}_1)\, \Psi_{n2}(\boldsymbol{k}_2)\, \Psi_{n3}(\boldsymbol{k}_3)\, \Psi_{n4}(\boldsymbol{k}_4) \tag{16.37}$$

representing the quartic term. Also, $\delta \equiv 1 - U\rho(0)$, and V is the system volume. This means that we can next employ a modified form of the renormalization group approach, now reformulated in $(\boldsymbol{k}, \omega)$ space. The treatment of the Mott metal–insulator transition within the same methodology is much more involved; no final answers have been provided as yet.[10]

In Table 16.1, we list examples of scaling for the physical quantities as $T \to 0$: FM and AFM refer to ferromagnetic and antiferromagnetic quantum transitions, respectively. The critical exponents as $T \to 0$ are either different from their classical values or even take the form of logarithmic singularities.

16.8 CONCLUDING REMARK

The story of quantum critical phenomena started with spin models, for example, the Ising model in transverse magnetic field[12] and then proliferated to fermionic

10. See, e.g., M. Continentino, *Quantum Scaling in Many-Body Systems* (Cambridge University Press, 2017).
11. Courtesy of D. Kaczorowski.
12. Cf. S. Sachdev, *Quantum Phase Transitions* (Cambridge University Press, 1999); *Understanding Quantum Phase Transitions*, edited by L.D. Carr (Taylor & Francis, Boca Raton, 2011); I. Herbut, *A Modern Approach to Critical Phenomena* (Cambridge University Press, 2007).

systems, where spectacular measurable effects can appear. The reader is advised to consult those references and review articles appearing under the headings of *Quantum Phase Transitions* and *Quantum Critical Phenomena*.

Chapter 17

Supplement: Going Beyond the Gaussian Formulation

We now take up the important problem of proceeding beyond the Gaussian approximation to generate the partition functions and correlation functions for the Landau–Ginzburg model. This will be done in two steps. The first is laborious but provides more practice in handling functionals. Those interested in an approach with easier access to the topic of Feynman diagrams may proceed directly to the second method, beginning ahead of Eq. (17.15).

17.1 METHOD 1: TWO-POINT CORRELATION FUNCTIONS

Consider again the partition functional that goes beyond the Gaussian model:

$$\begin{aligned}\mathcal{Z}[J] = \int \mathcal{D}\varphi e^{-\widehat{\mathcal{H}}[J]} = \int \mathcal{D}\varphi e^{-\widehat{\mathcal{H}}_0[J]} \exp\left[-\int d^d z \varphi^4/4!\right] \equiv \\ \mathcal{Z}_0[J] \exp\left[-\int d^d z \lambda_0 \varphi^4/4!\right].\end{aligned} \tag{17.1}$$

In Chapter 15, this expression was rewritten in the basic form

$$\begin{aligned}\mathcal{Z}[J] = \mathcal{Z}_0[0] \sum_{n=0}^{\infty} \frac{1}{n!} \left(-\frac{\lambda_0}{4!} \int d^d z \frac{\partial^4}{\partial J^4(z)}\right)^n e^{\mathcal{F}}, \\ \mathcal{F} \equiv \left[\frac{1}{2} \int d^d \boldsymbol{x} \int d^d \boldsymbol{y} J(\boldsymbol{x}) \Delta(\boldsymbol{x}-\boldsymbol{y}) J(\boldsymbol{y})\right].\end{aligned} \tag{17.2}$$

The *four-point correlation function* $G^{(n)}(\boldsymbol{x}_1, \ldots, \boldsymbol{x}_4) \equiv \langle \varphi(\boldsymbol{x}_1) \cdots \varphi(\boldsymbol{x}_4) \rangle$, by analogy to Eq. (3.19), is determined as follows:

$$G^{(n)}(\boldsymbol{x}_1, \ldots, \boldsymbol{x}_n) = \frac{1}{Z_0[J]} \frac{\delta}{\delta J(\boldsymbol{x}_1)} \cdots \frac{\delta}{\delta J(\boldsymbol{x}_4)} Z[J]. \tag{17.3}$$

The resulting zero-order relation for $Z[J]$ and the corresponding correlation function were exhibited in Chapter 15. To proceed with the first-order case, we consider the term $n = 1$ of Eq. (17.2). For use in Eq. (17.3),

A Primer to the Theory of Critical Phenomena. http://dx.doi.org/10.1016/B978-0-12-804685-2.00023-1

we must first execute the quadruple functional operation $\int \cdots \int d^d z_1 \cdots d^d z_4$ $[\delta/\delta J(z_4) \cdots \delta/\delta J(z_1)]$ on $e^{\mathcal{F}}$, to be followed by the double differentiation $\delta/\delta J(\boldsymbol{x}_1)\ \delta/\delta J(\boldsymbol{x}_2)$.

The initial operation involves

$$\begin{aligned} \frac{\delta e^{\mathcal{F}}}{\delta J(z_1)} &= \left[\frac{1}{2}\int d^d \boldsymbol{y} \Delta(z_1 - \boldsymbol{y}) J(\boldsymbol{y}) + \frac{1}{2}\int d^d \boldsymbol{x} \Delta(z_1 - \boldsymbol{x}) J(\boldsymbol{x})\right] e^{\mathcal{F}} \\ &= \left[\int d^d \boldsymbol{y} \Delta(z_1 - \boldsymbol{y}) J(\boldsymbol{y})\right] e^{\mathcal{F}}. \end{aligned} \tag{17.4}$$

The three subsequent operations follow exactly the same pattern, which is left as a straightforward, somewhat laborious, exercise. It emerges that the four-fold differentiation results in two sets of terms: three integrals of the type

$$I_1 \equiv \int \cdots \int d^d z_1 \cdots d^d z_4 \Delta(z_1 - z_2) \Delta(z_3 - z_4) e^{\mathcal{F}} \tag{17.5}$$

with the indices properly permuted, (1234) → (1324) → (1423), and six identical integrals of the type

$$\begin{aligned} I_2 \equiv \int \cdots \int d^d z_1 \cdots d^d z_4 \Delta(z_3 - z_4) \times \\ \int d^d \boldsymbol{y} \Delta(z_1 - \boldsymbol{y}) J(\boldsymbol{y}) \int d^d \boldsymbol{x} \Delta(z_2 - \boldsymbol{x}) J(\boldsymbol{x}) e^{\mathcal{F}} \end{aligned} \tag{17.6}$$

with the various indices permuted.

The remaining operations $\delta/\delta J(\boldsymbol{x}_1)\delta/\delta J(\boldsymbol{x}_2)$ are executed precisely as in Eqs. (15.25)–(15.28). Working in this manner on the three I_1, then setting $J = 0$, and introducing the neglected prefactor, produces

$$\begin{aligned} Q^{(2)}(\boldsymbol{x}_1, \boldsymbol{x}_2)\Big|_{11} = -\frac{\lambda_0}{8} \int \cdots \int d^d z_1 \cdots d^d z_4 \Delta(z_1 - z_2)\, \Delta(z_3 - z_4) \times \\ \Delta(\boldsymbol{x}_1 - \boldsymbol{x}_2), \end{aligned} \tag{17.7}$$

and setting $z_4 = z_3 = z_2 = z_1 \equiv z$, we finally obtain

$$\begin{aligned} Q^{(2)}(\boldsymbol{x}_1, \boldsymbol{x}_2)\Big|_{11} &= -\frac{\lambda_0}{8} \int d^d z [\Delta(z - z)]^2 \Delta(\boldsymbol{x}_1 - \boldsymbol{x}_2) \\ &= -\frac{\lambda_0}{8} \Delta^2(0) \Delta(\boldsymbol{x}_1 - \boldsymbol{x}_2) \int d^d z. \end{aligned} \tag{17.8}$$

Proceeding similarly with the six members of the second set, the first differentiation yields

$$Q^{(2)}(\boldsymbol{x}_1,\boldsymbol{x}_2)\Big|_{12}=\frac{\delta}{\delta J(\boldsymbol{x}_1)}\int\cdots\int d^d z_1\cdots d^d z_4$$
$$\times\left\{-\frac{\lambda_0}{4}\Delta(z_3-z_4)\left[\Delta(z_1-\boldsymbol{x}_2)\int d^d\boldsymbol{y}\Delta(z_2-\boldsymbol{y})J(\boldsymbol{y})\right.\right.$$
$$\left.\left.+\Delta(z_2-\boldsymbol{x}_2)\int d^d\boldsymbol{x}\Delta(z_1-\boldsymbol{x})J(\boldsymbol{x})\right]\right\}e^{\mathcal{F}}$$
$$\times\left[-\frac{\lambda_0}{4}\Delta(z_3-z_4)\int d^d\boldsymbol{y}\,\Delta(z_1-\boldsymbol{y})J(\boldsymbol{y})\right.$$
$$\left.\times\int d^d\boldsymbol{x}\Delta(z_2-\boldsymbol{x})J(\boldsymbol{x})\int d^d\boldsymbol{v}\,\Delta(\boldsymbol{x}_2-\boldsymbol{v})J(\boldsymbol{v})\right]e^{\mathcal{F}}. \tag{17.9}$$

A second differentiation yields

$$Q^{(2)}(\boldsymbol{x}_1,\boldsymbol{x}_2)\Big|_{12}=\int\cdots\int d^d z_1\cdots d^d z_4$$
$$\times\left[\begin{array}{c}\left\{-\frac{\lambda_0}{4}\Delta(z_3-z_4)\left[\Delta(z_1-\boldsymbol{x}_2)\Delta(z_2-\boldsymbol{x}_1)+\Delta(z_2-\boldsymbol{x}_2)\Delta(z_1-\boldsymbol{x}_1)\right]\right\}e^{\mathcal{F}}\\ +\text{many terms consisting of multiplying polynomials, each containing } J(\cdot),\\ \text{that vanish on setting } J=0.\end{array}\right] \tag{17.10}$$

Then set $z_4=z_3=z_2=z_1\equiv z$ and $J=0$ to obtain

$$Q^{(2)}(\boldsymbol{x}_1,\boldsymbol{x}_2)\Big|_{12}=-\frac{\lambda_0}{2}\int d^d z\Delta(\boldsymbol{x}_1-z)\,\Delta(z-z)\Delta(z-\boldsymbol{x}_2), \tag{17.11}$$

so that after assembly of the various pieces, we find that

$$G^{(2)}(\boldsymbol{x}_1,\boldsymbol{x}_2)=-\frac{Z_0[0]}{Z[J]}\left[\begin{array}{l}\Delta(\boldsymbol{x}_1-\boldsymbol{x}_2)-\frac{\lambda_0}{8}\int d^d z\,(\Delta(z-z))^2\,\Delta(\boldsymbol{x}_1-\boldsymbol{x}_2)\\ -\frac{\lambda_0}{2}\int d^d z\Delta(\boldsymbol{x}_1-z)\,\Delta(z-z)\Delta(z-\boldsymbol{x}_2)\end{array}\right]. \tag{17.12}$$

Next, we attend to the function $Z[J]$ in the denominator. According to Eq. (17.2), the zero-order contribution is simply $Z_0[0]$. The first-order term is derived as outlined by the discussion leading to Eq. (17.8), which must be multiplied by $Z_0[0]\lambda_0/8$. Accordingly, we find that

$$Z[J]=Z_0[0]\left\{1-(\lambda_0/8)\left[\int d^d z\,(\Delta(z-z))^2\,\Delta(\boldsymbol{x}_1-\boldsymbol{x}_2)\right]\right\}. \tag{17.13}$$

Inserting Eq. (17.13) into (17.12), expanding, and retaining terms only to first order in λ_0, we obtain

$$G^{(2)}(\boldsymbol{x}_1, \boldsymbol{x}_2) = \Delta(\boldsymbol{x}_1 - \boldsymbol{x}_2) - \frac{\lambda_0}{2} \int d^d z \Delta(\boldsymbol{x}_1 - z)\, \Delta(z - z)\, \Delta(z - \boldsymbol{x}_2), \tag{17.14}$$

which is the final result of interest. Notice the concatenation of Δ functions inside the integral; this also sets the pattern for higher-order terms: here the "exact" propagator up to first order has been expressed as a convolution of bare propagators.

17.2 METHOD 2: THE GENERATING FUNCTIONAL FOR ESTABLISHING CORRELATIONS FOR A SINGLE VARIABLE

We now turn to the second method for specifying the all-important n-point correlation function $\langle \varphi(\boldsymbol{x}_1)\, \varphi(\boldsymbol{x}_2) \cdots \varphi(\boldsymbol{x}_n) \rangle$; this also provides an easier access to the related topic of Feynman diagrams.

We begin with the case of a single variable, following the procedure of Le Bellac.[1] Consider the following functions in consonance with Eq. (15.22):

$$Z[J] = \int d\phi e^{-(1/2)A^{-1}\phi^2 + J\phi}, \quad Z[0] = \int d\phi e^{-(1/2)A^{-1}\phi^2} = \sqrt{2\pi A}, \tag{17.15}$$

which obviously relate to Gaussian distributions with a (fictitious) field J. Now introduce the Hubbard–Stratonovich transformation $\phi = \varphi + JA$, introduced earlier, whereby the exponent in Eq. (17.15) reads $-(1/2)\phi A^{-1}\phi + J\phi = -(1/2)\varphi A^{-1}\varphi + (1/2)JAJ$, whence Eq. (17.15) may be rewritten as

$$Z[J] = Z[0] e^{(1/2)[JAJ]}. \tag{17.16}$$

We now generalize the latter as follows: for an unnormalized probability distribution function $P(\varphi) \geq 0$ over a single variable function $\varphi(x)$, we define a *generating function* as

$$Z[J] = \int d\varphi P(\varphi) e^{J\varphi}. \tag{17.17}$$

Using Eq. (3.B.1) *ff* as templates, it is easy to note that by differentiation of the functional $Z[J]$ with respect to the fictional field J, we can obtain the expectation value of φ^r in the form

$$\langle \varphi^r \rangle \equiv \frac{\int d\varphi \varphi^r P(\varphi)}{\int d\varphi P(\varphi)} = \frac{1}{Z[0]} \left. \frac{\partial^r Z[J]}{\partial J^r} \right|_{J=0}. \tag{17.18}$$

1. Michel Le Bellac, *Quantum and Statistical Field Theory*, Clarendon Press, Oxford (2002), Secs. 5.1 and 5.2.

17.2.1 Extension to *n* Variables

The following will be based on use of the unnormalized Gaussian distribution function

$$P(\varphi) = \exp\left(-\frac{1}{2}\varphi\frac{1}{A}\varphi\right). \tag{17.19}$$

We can immediately generalize it to the case of n variables $\varphi(x_i) \equiv \varphi_i$, $i = 1, \ldots, n$, together with the associated unnormalized Gaussian-type distribution function $P(\varphi_1, \ldots, \varphi_n) = \exp\left(-1/2\boldsymbol{\varphi}^T \boldsymbol{A}^{-1}\boldsymbol{\varphi}\right)$, where $\boldsymbol{\varphi}^T \equiv (\varphi_1, \ldots, \varphi_n)$, $\boldsymbol{\varphi}$ is the corresponding column vector, and $\boldsymbol{A}$ is a matrix with A_{ij} as entries. Thus,

$$\boldsymbol{\varphi}^T \boldsymbol{A}^{-1}\boldsymbol{\varphi} \equiv \sum_{i=1}^{n}\sum_{j=1}^{n} \varphi_i A_{ij}^{-1}\varphi_j. \tag{17.20}$$

As a generalization of Eq. (17.15), we generate partition functionals of the form

$$Z[\boldsymbol{J}] = \int \cdots \int \prod_{i=1}^{n} d\phi_i \exp\left(-(1/2)\boldsymbol{\phi}^T \boldsymbol{A}^{-1}\boldsymbol{\phi} + \boldsymbol{J}^T\boldsymbol{\phi}\right), \quad \boldsymbol{J}^T\boldsymbol{\phi} \equiv \sum_{i=1}^{n} J_i\phi_i, \tag{17.21}$$

and introduce the change of variable: $\boldsymbol{\phi} = \boldsymbol{\varphi} + \boldsymbol{A}\boldsymbol{J}$, whereby the exponent reads

$$-\frac{1}{2}\boldsymbol{\phi}^T \boldsymbol{A}^{-1}\boldsymbol{\phi} + \boldsymbol{J}^T\boldsymbol{\phi} = -\frac{1}{2}\boldsymbol{\varphi}^T \boldsymbol{A}^{-1}\boldsymbol{\varphi} + \frac{1}{2}\boldsymbol{J}^T\boldsymbol{A}\boldsymbol{J}. \tag{17.22}$$

The generating function thus assumes the form

$$Z[\boldsymbol{J}] = Z_0[0]\exp\left(\frac{1}{2}\boldsymbol{J}^T\boldsymbol{A}\boldsymbol{J}\right), \quad Z_0[0] \equiv \int \cdots \int \prod_{i=1}^{N} d\varphi_i \exp\left(-\frac{1}{2}\boldsymbol{\varphi}^T \boldsymbol{A}^{-1}\boldsymbol{\varphi}\right). \tag{17.23}$$

The above substitution converts the integral (17.21), which involves both $\boldsymbol{\phi}^T\boldsymbol{A}^{-1}\boldsymbol{\phi}$ and $\boldsymbol{J}^T\boldsymbol{\phi}$, to a Gaussian-type integral that features $\boldsymbol{\varphi}^T\boldsymbol{A}^{-1}\boldsymbol{\varphi}$, multiplied by the factor $\exp\left(\frac{1}{2}\boldsymbol{J}^T\boldsymbol{A}\boldsymbol{J}\right)$, which now serves as the operand of interest.

This leads us directly to the formation of higher-order correlation functions by differentiation as shown earlier: we temporarily exclude the factor $Z_0[0]$ and use Eq. (17.23) as operand in a field-free region to consider

$$\begin{aligned}\langle\varphi_{i1}\ldots\varphi_{il}\ldots\varphi_{i2n}\rangle_0 &\sim \left.\frac{\delta^{2n}}{\delta J_{i1}\cdots\delta J_{il}\cdots\delta J_{i2n}} Z[\boldsymbol{J}]\right|_{\boldsymbol{J}=0} \\ &= \left.\frac{\delta^{2n}}{\delta J_{i1}\cdots\delta J_{il}\cdots\delta J_{i2n}}\frac{1}{n!}\frac{1}{2^n}\left(\boldsymbol{J}^T\boldsymbol{A}\boldsymbol{J}\right)^n\right|_{\boldsymbol{J}=0}.\end{aligned} \tag{17.24}$$

This calls for several comments: The exponential function $Z[\boldsymbol{J}]$ was first expanded in a power series in $\left(\frac{1}{2}\boldsymbol{J}^T\boldsymbol{A}\boldsymbol{J}\right)^l$ for which the operand on the right of Eq. (17.24) is the nth term. It alone has survived the $2n$-fold differentiations on $Z[\boldsymbol{J}]|_{J=0}$ since all terms with $l < n$ being differentiated in Eq. (17.20) automatically vanish. Terms with $l > n$ yield residual multiplicative factors containing J; all these terms vanish on setting $J = 0$. As shown by analogy to Eq. (17.20), the surviving operand consists of a double summation raised to the power n. The requisite differentiations become extremely unwieldy for terms with $n > 2$; other approaches will be shortly introduced to address this problem. The powers of J^2 in the operand require $2n$-fold differentiations to generate nonvanishing correlations. When an odd number of differentiations is attempted, there always remain terms in J that with $J = 0$ cause such a correlation function to vanish.

17.2.2 Contractions and Wick's Theorem

The requisite differentiations in Eq. (17.24) can be greatly simplified. Consider the following second moment:

$$\langle\varphi_{i1}\varphi_{i2}\rangle_0 = \frac{\delta^2}{\delta J_{i1}\delta J_{i2}}\left(\frac{1}{2}\sum_k\sum_l J_k A_{kl} J_l\right) = A_{i1i2} \equiv \overbrace{\varphi_{i1}\varphi_{i2}}\,. \qquad (17.25)$$

As shown, A_{i1i2} is represented by the second Gaussian moment, as indicated on the left, and is rewritten on the right in a form termed a *contraction*, which involves φ_{i1} and φ_{i2}. When the contraction scheme is applied to the multiple differentiations of Eq. (17.24), then $(2n)!$ terms are generated, as is made clear by studying the operations in the Appendix. Dividing by the numerical factor $1/n!2^n$ present in Eq. (17.24), we obtain $(2n-1)!! = \cdots 7\cdot 5\cdot 3\cdot 1$ paired terms, a factorial representation that involves solely odd integers. But this is precisely the number of $\varphi_{il}\varphi_{im}$ contraction pairs that can be generated from the differentiations of $\langle\varphi_{i1}\ldots\varphi_{il}\ldots\varphi_{i2n}\rangle_0$, since in the pairing process, you start with $(2n-1)$ choices for the first pair, $(2n-3)$ remaining choices for the second selection, and so on. The upshot of the above is that the correlation function $\langle\varphi_{i1}\cdots\varphi_{il}\cdots\varphi_{i2n}\rangle_0$ can always be decomposed into sums of products of contractions $\langle\varphi_{i1}\varphi_{i2}\rangle_0$, which enormously simplifies subsequent operations. This feature is known as *Wick's theorem*.

To illustrate the above process, consider correlations involving four order parameters, setting $n = 2$; for consistency with our earlier formulation of Eqs. (12.31)–(12.35), we now replace $\boldsymbol{A}$ with $\Delta(\boldsymbol{x}-\boldsymbol{y})$. The object of interest is

$$\langle\varphi_1\ldots\varphi_4\rangle_0 = \frac{\delta^4}{\delta J_1\ldots\delta J_4}\frac{1}{8}\left(\boldsymbol{J}^T\Delta\boldsymbol{J}\right)^2\Big|_{J=0}. \qquad (17.26)$$

Without invoking Wick's theorem, the required brute force operations are shown in mind-numbing detail in the Appendix. We obtain the following result, exclusive of multiplicative numerical factors:

$$\begin{aligned}\langle\varphi_1\varphi_2\varphi_3\varphi_4\rangle_0 = {} & \Delta(\boldsymbol{x}_1-\boldsymbol{x}_2)\,\Delta(\boldsymbol{x}_3-\boldsymbol{x}_4)+\Delta(\boldsymbol{x}_1-\boldsymbol{x}_3)\,\Delta(\boldsymbol{x}_2-\boldsymbol{x}_4) \\ & +\Delta(\boldsymbol{x}_1-\boldsymbol{x}_4)\,\Delta(\boldsymbol{x}_2-\boldsymbol{x}_3)\,.\end{aligned} \tag{17.27}$$

This is entirely consistent with the Wick formulation

$$\begin{aligned}\langle\varphi_1\varphi_2\varphi_3\varphi_4\rangle_0 = {} & \langle\varphi_1\varphi_2\rangle_0\,\langle\varphi_3\varphi_4\rangle_0+\langle\varphi_1\varphi_3\rangle_0\,\langle\varphi_2\varphi_4\rangle_0 \\ & +\langle\varphi_1\varphi_4\rangle_0\langle\varphi_2\varphi_3\rangle_0.\end{aligned} \tag{17.28}$$

As a reminder, in Eq. (12.31), we linked the delta function to the two-point correlation function based on Gaussians. The above fourfold correlation function has thus been decomposed into three pairs of two-point contractions, with all possible pair permutations of the indices. Quite generally, Wick's theorem states that any even multiple moments of Gaussian distributions can be decomposed into sets of second moments, containing all possible permutations of the order parameters. Odd multiple moment terms vanish.

17.2.3 Correlation Functions Involving a Single Variable

Let us now determine the correlation function that corresponds to the use of the distribution

$$P(\varphi)=\exp\left(-\frac{1}{2}\varphi\frac{1}{A}\varphi-\frac{\lambda_0}{4!}\varphi^4\right). \tag{17.29}$$

Expanding the second factor to first order, we find that the corresponding partition function assumes the form

$$Z[0]=\int d\varphi\, e^{-(1/2)A^{-1}\varphi^2}\left(1-\lambda_0\varphi^4/4!\right)=\sqrt{2\pi A}\left(1-\lambda_0A^2/8\right). \tag{17.30}$$

The result on the right is derived in the Appendix. This result will be inserted into the denominator of Eq. (17.18).

The numerator of Eq. (17.18) is formed via (see the Appendix)

$$\int d\varphi\,\varphi^2 e^{-(1/2)A^{-1}\varphi^2}\left(1-\lambda_0\varphi^4/4!\right)=\sqrt{2\pi A}\left(A-5\lambda_0A^3/8\right). \tag{17.31}$$

Accordingly, the second moment is found by division of Eq. (17.31) by (17.30). To first order in λ_0, we obtain

$$\langle\varphi^2\rangle=A\left(1-\lambda_0A^2/2\right). \tag{17.32}$$

Note that the factor $\sqrt{2\pi A}$ has canceled from the ratio; the constant corresponding to the Gaussian integral is irrelevant in the formation of the correlation function.

17.2.4 Determination of Ginzburg–Landau-type Correlation Functions for Several Variables; Brief Introduction to Feynman Diagrams

Let us next study, to first order in λ_0, the $G^{(2)}(\boldsymbol{x}-\boldsymbol{y})$ correlation function, as determined by the first-order integral

$$I(\boldsymbol{x},\boldsymbol{y}) = \int D\varphi\,\varphi(\boldsymbol{x})\varphi(\boldsymbol{y})e^{-\hat{H}} \approx \int D\varphi\,\varphi(\boldsymbol{x})\varphi(\boldsymbol{y})e^{-\hat{H}_0}\left[1-\frac{\lambda_0}{4!}\int d^d z\varphi^4(z)\right]. \tag{17.33}$$

The first term in square brackets yields (see Eqs. (12.34) and (12.35))

$$N\langle\varphi(\boldsymbol{x})\,\varphi(\boldsymbol{y})\rangle_0 = NG_0(\boldsymbol{x}-\boldsymbol{y}),\quad N\equiv\int D\varphi\,e^{-\hat{H}_0} = Z_0(0)\,. \tag{17.34}$$

To deal with the second term, with neglect of unneeded constants, namely

$$\int\left[D\varphi\,\varphi(\boldsymbol{x})\varphi(\boldsymbol{y})e^{-\hat{H}_0}\int\cdots\int\prod_{i=1}^{4}d^d z_i\,\{\varphi(z_1)\,\varphi(z_2)\,\varphi(z_3)\,\varphi(z_4)\}\right], \tag{17.35}$$

we can now apply Wick's theorem, so as to generate all possible contractions consistent with Eq. (17.35). Clearly, the contemplation and handling of all these integrals are taxing; here a pictorial representation of Wick's theorem comes to the rescue, based on *Feynman diagrams*, in the following manner: We label x and y as "external points," whereas the z_i position variables that get integrated out are designated as "internal points" or "vertices." Let us represent the four vertices by four vertically aligned points in Fig. 17.1 and the external points by the horizontal alignment of crosses labeled x and y.

The Wick constructions are executed symbolically by drawing all possible connections among the points and crosses. Two possible sets of diagrams are encountered as shown in Fig. 17.2. Part (A) represents one of 12 possibilities that involve the connection of the external points $\boldsymbol{x}$ and $\boldsymbol{y}$ to two out of four vertices: $\boldsymbol{x}$ can be connected to any of the four z vertices, and $\boldsymbol{y}$ to any of the three remaining vertices; an additional link is then needed to connect the remaining two vertices. This set represents a collection of *connected diagrams.* Check yourself by drawing all such possible configurations. Part (B) represents one of three possible configurations in which x is connected to y, and the four

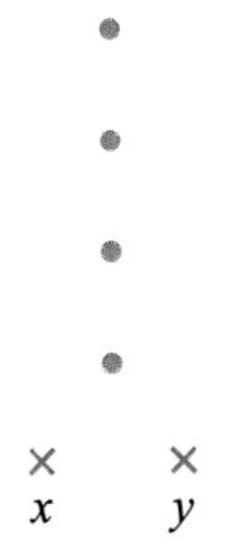

FIGURE 17.1 Pictorial representation of vertices (heavy dots) and external points (crosses) for use in Feynman diagrams.

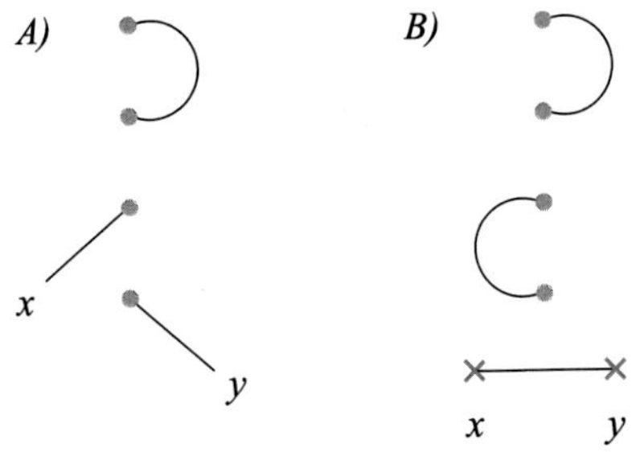

FIGURE 17.2 Pictorial representation of contractions constructed from integrals (17.35) as Feynman diagrams. Contractions are indicated by connections. Parts (A) and (B) show two distinct types of connections.

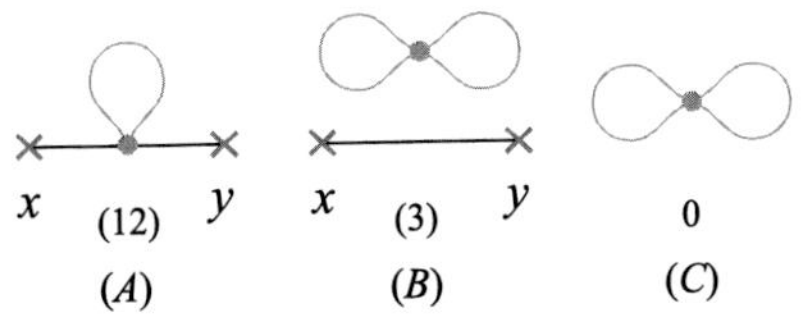

FIGURE 17.3 Condensed versions of Fig. 17.2, as obtained by melding of the vertices to generate the Feynman diagrams. Part (C) represents a vacuum bubble.

vertices are linked pairwise. This leads to a separation of configurations, leading to *disconnected diagrams*. We encounter a total of $(6-1)!! = 15$ possibilities.

An economy in representations is achieved by connecting and melding the four vertices into a single point, thereby generating loops and compressing the original diagrams (of which Figs. 17.2A and B are two examples) as shown in Fig. 17.3. Associated with each diagram is the number of individual connections represented in the compressed version. We must remember that each vertex (closed dot in Fig. 17.3) really represents an integral over the z coordinate.

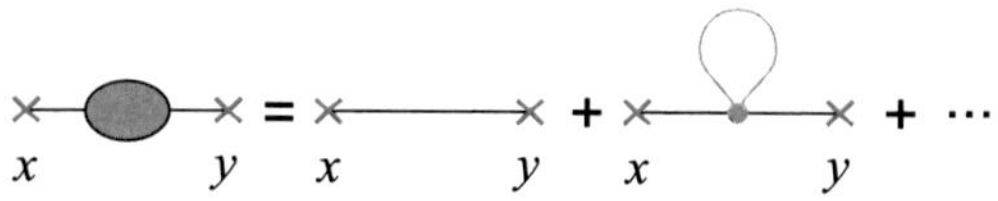

FIGURE 17.4 Pictorial representation via Feynman diagrams of Eq. (17.38).

17.2.5 Illustration of the Use of Feynman Diagrams

We now show how Feynman diagrams may be utilized; for ease of comparison with the prevailing literature, we replace $\Delta(\boldsymbol{x}_1 - \boldsymbol{x}_2)$ by $G_0(\boldsymbol{x}_1 - \boldsymbol{x}_2)$. Translated in reverse, the twelve contractions of type 17.3 (a) correspond to the integral $\int d^d z G_0(\boldsymbol{x}-\boldsymbol{z})G_0(0)G_0(\boldsymbol{z}-\boldsymbol{y})$, and the three contractions of type 17.3 (b) correspond to the integral $\int d^d z G_0(\boldsymbol{x}-\boldsymbol{y})G_0^2(0)$. The Wick decomposition of Eq. (17.35) as depicted in Fig. 17.3 thus gives rise to the following result: To first order in λ_0, the symmetry numbers for the two cases correspond to $12/4! = 1/2$ and to $3/4! = 1/8$, respectively. Thus, Figs. 17.3A, B, C correspond to

$$I(\boldsymbol{x},\boldsymbol{y}) = N\left[G_0(\boldsymbol{x}-\boldsymbol{y}) - \frac{\lambda_0}{2}\int d^d z G_0(\boldsymbol{x}-\boldsymbol{z})G_0(0)G_0(\boldsymbol{z}-\boldsymbol{y}) - \frac{\lambda_0}{8}G_0(\boldsymbol{x}-\boldsymbol{y})G_0^2(0)\int d^d z\right]. \tag{17.36}$$

To obtain the correlation function, we must divide by

$$Z[0] = \int D\varphi\, e^{-\hat{H}_0}\left[1 - \frac{\lambda_0}{4!}\int d^d z \varphi^4(z)\right] = N\left[1 - \frac{\lambda_0}{8}G_0^2(0)\int d^d z\right], \tag{17.37}$$

where the second term is represented by Fig. 17.3C. Then the correlation function to first order in λ_0 is given by

$$G^{(2)}(\boldsymbol{x}-\boldsymbol{y}) = G_0(\boldsymbol{x}-\boldsymbol{y}) - \frac{\lambda_0}{2}\int d^d z G_0(\boldsymbol{x}-\boldsymbol{z})G_0(0)\,G_0(\boldsymbol{z}-\boldsymbol{y}), \tag{17.38}$$

which is in agreement with our earlier result.

Notice how deftly by use of Wick's theorem the original hexafunctional integrations have been handled. As a result, the correlation function, or *propagator*, on the left has been expanded in terms of Gaussian correlations on the right. The graphical representation of Eq. (17.38) is provided in Fig. 17.4. Notice also the absence of the disconnected diagram. The discarded "infinity" loop is called a *vacuum fluctuation*, a term borrowed from quantum field theory. It corresponds to the elimination of terms in $G_0^2(0)$ on carrying out the above-mentioned division. This turns out to be a general feature of perturbative expansions.

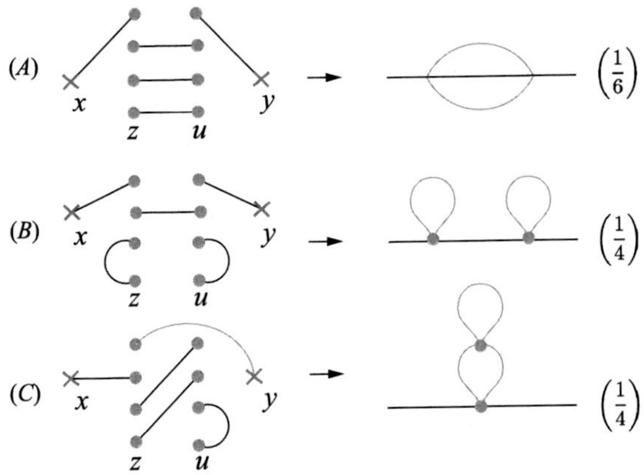

FIGURE 17.5 Pictorial Representation via diagrams of the second-order expansion of the four-point correlation function. Parts (A)–(C) indicate three distinct types of diagrams; on the right the corresponding Feynman diagrams and symmetry numbers are shown. Consult text for explanations.

17.2.6 Second-order Expansion

Let us now briefly refer to the second-order expansion of the four-point correlation function, which calls for a study of expressions of the type

$$\mathrm{I} = \left\langle \varphi(\boldsymbol{x})\varphi(\boldsymbol{y})\frac{\lambda_0^2}{2!}\cdot\frac{1}{4!}\int d^d z\varphi^4(z)\frac{1}{4!}\int d^d \boldsymbol{u}\varphi^4(\boldsymbol{u})\right\rangle_0 . \tag{17.39}$$

We deal with this expression via Feynman diagrams: Separate out the external points $\boldsymbol{x}$ and $\boldsymbol{y}$ and line up the four z_i and $\boldsymbol{u}_j$ in two vertical columns. Fig. 17.5A displays one possible interconnection for the contractions in Wick's theorem. This represents the case where the external point $\boldsymbol{x}$ can be connected to any of the eight vertices z_i or $\boldsymbol{u}_j$; $\boldsymbol{y}$ is then connected to any of the four vertices in the column not containing the $\boldsymbol{x}$ connection. (For $\boldsymbol{x}$ and $\boldsymbol{y}$ connections that end up in the same column, see below). We then draw all possible $z - \boldsymbol{u}$ links across unconnected locations in the two columns (for $z - z$ and $\boldsymbol{u} - \boldsymbol{u}$ pairings, see below). The various possible configurations involve both uncrossed and crossing linkages, as is shown by the examples in Fig. 17.6A. As you need to check, six distinct bond patterns can be generated under the above restrictions; Fig. 17.5A is simply the most symmetric. The Feynman diagram corresponding to this set is shown on the right. The symmetry factor is $8 \times 4 \times 6/(2! \times 4! \times 4!) = 1/6$.

Two other patterns can be similarly constructed: the first of these $\boldsymbol{x}$ and $\boldsymbol{y}$ are connected as before, but only one cross-linked connection is established between z and $\boldsymbol{u}$ vertices, forcing one $z - z$ and one $\boldsymbol{u} - \boldsymbol{u}$ link to be introduced. You need to verify that nine distinct configurations of this type arise under these conditions. The most symmetric configuration is shown in

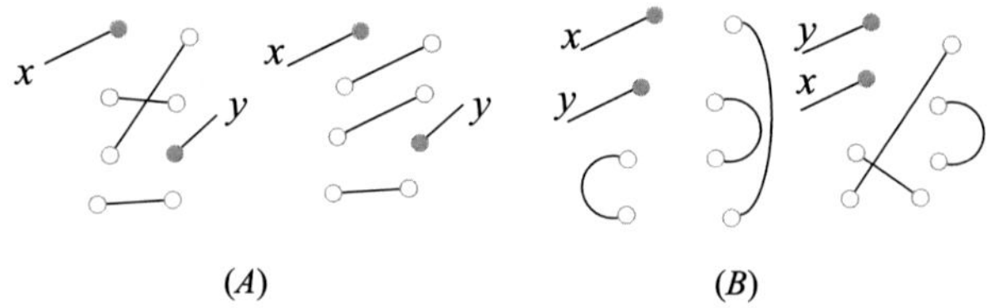

FIGURE 17.6 Selected diagrams used in the construction of Fig. 17.5 as explained in the text.

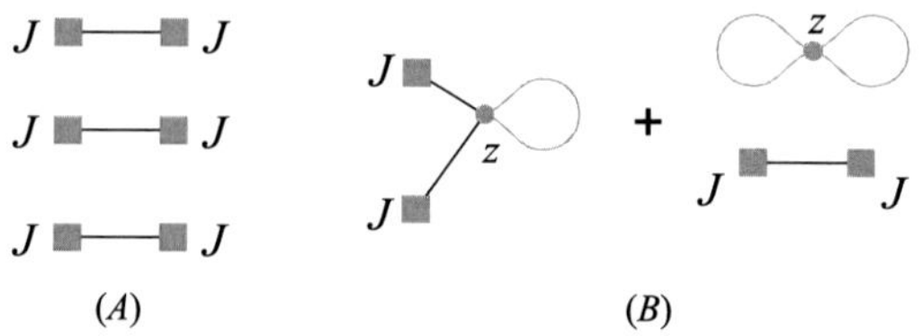

FIGURE 17.7 Diagrams generated by execution of Exercise 17.3.

Fig. 17.5B, along with the condensed version and with the symmetry number $8 \times 4 \times 9/(2! \times 4! \times 4!) = 1/4$. In the third set, Fig. 17.5C, $\boldsymbol{x}$ and $\boldsymbol{y}$ are joined to two vertices in either the $\boldsymbol{z}$ or the $\boldsymbol{u}$ column. If a third connection is formed between the unoccupied vertices in the same column (say $\boldsymbol{z}$), the remaining links occupy the opposite column (say $\boldsymbol{u}$). When allowing for an interchange of $\boldsymbol{x}$ with $\boldsymbol{y}$, we generate a pattern of 12 units, of which Fig. 17.6B is an example. On the other hand, placing the third link in the alternate column mandates that two additional connections, either nonintersecting or intersecting, be constructed across the columns, for a total of 24, of which one of the most symmetric is shown in Fig. 17.7. The condensed version for the entire third set is shown in Fig. 17.5C, together with its symmetry number $\frac{1}{4}$.

Note that additional linkages arise by connecting $\boldsymbol{x}$ and $\boldsymbol{y}$ directly; these are only of pedantic interest, since it can be shown that they do not contribute to correlation functions that are, after all, the principal items of interest.

We can now work with the Figs. 17.4, 17.5, and 17.6 in reverse to generate the mathematical expression for the second-order contribution to the four-point correlation function, yielding

$$G_0^{(4,2)} = \frac{\lambda_0^2}{2!}\left[\begin{array}{l} \frac{1}{6}\int d^d\boldsymbol{z}\int d^d\boldsymbol{u}\, G_0(\boldsymbol{x}-\boldsymbol{z})G_0^3(\boldsymbol{z}-\boldsymbol{u})G_0(\boldsymbol{u}-\boldsymbol{y}) \\ +\frac{1}{4}\int d^d\boldsymbol{z}\int d^d\boldsymbol{u}\, G_0(\boldsymbol{x}-\boldsymbol{z})G_0(0)G_0(\boldsymbol{z}-\boldsymbol{u})G_0(0)G_0(\boldsymbol{u}-\boldsymbol{y}) \end{array}\right]$$
$$+\frac{\lambda_0^2}{2!}\left[\begin{array}{l} \frac{1}{12}\int d^d\boldsymbol{z}\int d^d\boldsymbol{u}\, G_0(\boldsymbol{x}-\boldsymbol{z})G_0^3(0)G_0(\boldsymbol{z}-\boldsymbol{y}) \\ +\frac{1}{6}\int d^d\boldsymbol{z}\int d^d\boldsymbol{u}\, G_0(\boldsymbol{x}-\boldsymbol{z})G_0(0)G_0(\boldsymbol{z}-\boldsymbol{u})G_0(\boldsymbol{u}-\boldsymbol{z})G_0(\boldsymbol{z}-\boldsymbol{y}) \end{array}\right]. \tag{17.40}$$

17.2.7 Summary

Without going into further detail, the above examples suggest the following general rules to set up a Feynman dictionary applicable to direct space, as exemplified by the above figures:

1. Every internal point (vertex) is assigned a factor $-\lambda_0$.
2. Every line $\boldsymbol{x}-\boldsymbol{y}$ is associated with the factor $G_0(\boldsymbol{x}-\boldsymbol{y})$, also known as a *propagator*.
3. Every point z_i involves an integral $\int d^d z_i$.
4. Every graph is associated with a symmetry factor. General, somewhat complex, rules have been developed for their specification, but the numbers can usually be determined by inspection of the diagrams.

For completeness, it should be mentioned that higher-order expansions are ordinarily carried out in reciprocal space, with its own rules for constructing Feynman diagrams.

17.2.8 Postscript

This is a far as we can go without resorting to more esoteric mathematics and diagrammatic techniques. To help in this task, some recommended monographs on the subject are listed separately.

17.2.9 Appendix

Part 1. We show here in detail the type of operations required to carry out the construction of correlation functions via Eq. (17.26) for $n = 2$. Excluding irrelevant numerical factors and parameters, the required operations involve as operand the function $(1/2)[\Delta_{11}J_1^2 + \Delta_{22}J_2^2 + \Delta_{33}J_3^2 + \Delta_{44}J_4^2] + (1/2)[J_1\Delta_{12}J_2 + J_1\Delta_{13}J_3 + J_1\Delta_{14}J_4] + (1/2)[J_2\Delta_{21}J_1 + J_2\Delta_{23}J_3 + J_2\Delta_{24}J_4] + (1/2)[J_3\Delta_{31}J_1 + J_3\Delta_{32}J_2 + J_3\Delta_{34}J_4] + (1/2)[J_4\Delta_{41}J_1 + J_4\Delta_{42}J_2 + J_4\Delta_{43}J_3] +$ the same terms with the delta indices reversed.

Since the above quantities are symmetric in the indices, the required derivative operations may be written out in the format

$$\int \prod dz \left(\frac{\delta}{\delta J_1} \frac{\delta}{\delta J_2} \frac{\delta}{\delta J_3} \frac{\delta}{\delta J_4} \left\{ \left[\frac{1}{2} \left(\Delta_{11} J_1^2 + \Delta_{22} J_2^2 + \Delta_{33} J_3^2 + \Delta_{44} J_4^2 \right) \right.\right.\right.$$
$$+ (J_1\Delta_{12}J_2 + J_1\Delta_{13}J_3 + J_1\Delta_{14}J_4)$$
$$\left.\left.\left. + (J_2\Delta_{23}J_3 + J_2\Delta_{24}J_4) + (J_3\Delta_{34}J_4) \right] \right\}^2 \right) \Bigg|_{J=0}$$
$$\equiv \int \prod dz \left(\frac{\delta}{\delta J_1} \frac{\delta}{\delta J_2} \frac{\delta}{\delta J_3} \frac{\delta}{\delta J_4} \left\{ \cdots \right\}^2 \Bigg|_{J=0} \right). \tag{17.41}$$

The first differentiation with $\delta J_l/\delta J_m = \delta_{lm}$, yields

$$\int \prod dz \left[2\frac{\delta}{\delta J_1}\frac{\delta}{\delta J_2}\frac{\delta}{\delta J_3}\left\{\cdots\right\}(\Delta_{14}J_1 + \Delta_{24}J_2 + \Delta_{34}J_3 + \Delta_{44}J_4)\right], \tag{17.42}$$

followed by

$$\int \prod dz \left\{ 2\frac{\delta}{\delta J_1}\frac{\delta}{\delta J_2}[(\Delta_{13}J_1 + \Delta_{23}J_2 + \Delta_{33}J_3 + \Delta_{34}J_4)(\Delta_{14}J_1 + \Delta_{24}J_2 + \Delta_{34}J_3 + \Delta_{44}J_4) + [\{\cdots\}](\Delta_{34})] \right\}, \tag{17.43}$$

and by

$$\int \prod dz \left\{ 2\frac{\delta}{\delta J_1}[(\Delta_{23})(\Delta_{14}J_1 + \Delta_{24}J_2 + \Delta_{34}J_3 + \Delta_{44}J_4) + (\Delta_{24})(\Delta_{13}J_1 + \Delta_{23}J_2 + \Delta_{33}J_3 + \Delta_{34}J_4) + (\Delta_{34})(\Delta_{12}J_1 + \Delta_{22}J_2 + \Delta_{23}J_3 + \Delta_{24}J_4)] \right\}. \tag{17.44}$$

Finally, after executing the last differentiation, we obtain the result cited in Eq. (17.27).

Part 2. The first term in Eq. (17.30) is clearly a Gaussian integral converging on $\sqrt{2\pi A}$. The second involves the integral $(-\lambda_0/4!)\int d\varphi\,\varphi^3\left[\varphi e^{-\varphi^2/2A}\right]$, which can be integrated by parts to yield $-\varphi^3 A e^{-\varphi^2/2A} + (3\lambda_0 A/4!) \times \int d\varphi\varphi\left(\varphi e^{-\varphi^2/2A}\right)$; the first term, being an odd function in φ, cancels out for any symmetrically disposed end points. The second term can be again integrated by parts to yield $-(\lambda_0 A^2/8)\sqrt{2\pi A}$.

The terms in (17.33) are handled similarly. Integration by parts of the first term repeats the last step above. The second term involves $\int d\varphi\,\varphi^3\left[\varphi e^{-\varphi^2/2A}\right] = -\varphi^5 A e^{-\varphi^2/2A} - 5A\int d\varphi\,\varphi^3\left[\varphi e^{-\varphi^2/2A}\right]$; the first term drops out, and the second has been dealt with above.

17.3 EXERCISES

Exercise 17.1. This exercise relates to a further study of partition functions. Verify the following relation:

$$Z[J] = -\lambda_0 \left\{ \begin{array}{l} \frac{1}{8}\int d^d z_0 \Delta^2(0) + \frac{1}{4}\iiint d^d z_0 d^d z_1 d^d z_2 \Delta(0)\Delta(z_1 - z_0) J(z_1) \\ \times \Delta(z_2 - z_0) J(z_2) + \frac{1}{4!}\iiint\iint d^d z_0 d^d z_1 d^d z_2 d^d z_3 d^d z_4 \\ \times \Delta(z_1 - z_0) J(z_1) \Delta(z_2 - z_0) J(z_2) \\ \times \Delta(z_3 - z_0) J(z_3) \Delta(z_4 - z_0) J(z_4) \end{array} \right\} e^{\mathcal{F}}, \tag{17.45}$$

which represents the first-order expansion of Eq. (17.2), exclusive of the factor $Z_0[0]$.

You need to execute the operations called for and use the definition, Eq. (12.32) for Δ, and the relations $\delta(\omega) = (2\pi)^{-1}\int_{-\infty}^{\infty} dt e^{-i\omega t}$ and $\int dt f(t)\delta(t - t_0) = f(t_0)$. The factor $(2\pi)^{-1}$ may be disregarded, as explained in conjunction with Eq. (12.24). Lastly, at the end you will have to insert the factor $e^{ikz_0} \times e^{-ikz_0}$ and exchange the integration variables $\int d^d z_0$ and $\int d^d z_n$ (convince yourself that this is permissible!) to check out the last two terms in the above equation.

Exercise 17.2. Earlier, we had tacitly assumed that the relation $\Delta_{ij} = \Delta_{ji}$ is satisfied. This is not obvious through a cursory inspection. Prove that this assumption is correct. Hint: Consult Appendix 14.B.

Exercise 17.3. For additional practice, consider the third-order term in the expansion of Eq. (17.2):

$$J_3 = \frac{1}{3!}\left[\iint \frac{1}{2} d^d z_1 d^d z_2 J(z_1) \Delta(z_2 - z_1) J(z_2)\right]^3 . \tag{17.46}$$

Note and verify the translation of the above into a Feynman diagram as shown in Fig. 17.7A. Generate the partition function via a fourfold differentiation $\delta^4/\delta J^4(z)$. To obtain a nonzero result, at least four of the J functions must have vertex z as their arguments, causing them to pile on top of each other at z. Then they are removed by the differentiation, leaving just z as the integration variable and as one of the arguments of the Δ function. Correspondingly, four links will have to emanate from that vertex. Verify the resulting diagrams as depicted in Fig. 17.7B.

Represent pictorially the additional two differentiations required to generate the correlation function. Verify that this leads to two diagrams almost identical to Fig. 17.7B, except that the Js are replaced by the coordinates $\boldsymbol{x}$ and $\boldsymbol{y}$. Then generate the corresponding mathematical formulas by carrying out the Feynman construction in reverse.

Bibliography

Suggestions for Further Reading

[1] D.J. Amit, Field Theory, the Renormalization Group, and Critical Phenomena, revised 2nd ed., World Scientific, Singapore, 1984.
[2] J.J. Binney, N.J. Dowrick, A.J. Fisher, M.E.J. Newman, The Theory of Critical Phenomena, Oxford University Press, Oxford, 1992.
[3] J. Cardy, Scaling and Renormalization in Statistical Physics, Cambridge University Press, 1996.
[4] C. di Castro, R. Raimondi, Statistical Mechanics and Applications to Condensed Matter, Cambridge University Press, 2015, Chapters 14 and 16–17.
[5] C. Domb, The Critical Point: A Historical Introduction to the Modern Theory of Critical Phenomena, Taylor & Francis, London, 1996, Chapters 6–7.
[6] N. Goldenfeld, Lectures on Phase Transitions and the Renormalization Group, Perseus Books, Reading, MA, 1992.
[7] I. Herbut, A Modern Approach to Critical Phenomena, Cambridge University Press, Cambridge, 2007.
[8] L.P. Kadanoff, Statistical Physics: Statics, Dynamics, Renormalization, Word Scientific, Singapore, 2000, Chapters 13–14.
[9] M. Kardar, Statistical Physics of Fields, Cambridge University Press, 2007.
[10] L.D. Landau, E.M. Lifshitz, Statistical Physics, 3rd edition, Elsevier, Amsterdam, 1980, Part 1, Chapter XIV.
[11] M. Le Bellac, Quantum and Statistical Field Theory, Clarendon Press, Oxford, 1991.
[12] S.-k. Ma, Modern Theory of Critical Phenomena, Westview Press, Oxford, 2000.
[13] W.D. McComb, Renormalization Methods, Clarendon Press, Oxford, 2004.
[14] H. Nishimori, G. Ortiz, Elements of Phase Transitions and Critical Phenomena, Oxford University Press, Oxford, 2011.
[15] P. Pfeuty, G. Toulouse, Introduction to the Renormalization Group and to Critical Phenomena, Wiley, New York, 1977.
[16] S. Sachdev, Quantum Phase Transitions, Cambridge University Press, Cambridge, 1999.
[17] H.E. Stanley, Introduction to Phase Transitions and Critical Phenomena, Oxford University Press, 1971.
[18] Y.M. Yeomans, Statistical Mechanics of Phase Transitions, Oxford University Press, 1992.
[19] J. Zinn-Justin, Quantum Field Theory and Critical Phenomena, Oxford University Press, 2004.
[20] J. Zinn-Justin, Phase Transitions and Renormalization Group, Oxford University Press, Oxford, 2007.

Index

L

M

N

O

P

Q

R

Printed in the United States
By Bookmasters